Behavior
of Captive
Wild
Animals

Contributing authors:

Linda P. Brett
Carl D. Cheney
Susan M. Essock
John C. Fentress
Rebecca Field
John Garcia
Carl R. Gustavson
Daniel J. Kelly
Hal Markowitz
William A. Myers
Heather Parr
Duane M. Rumbaugh
Ronald J. Schusterman
Victor J. Stevens
Gail Woodworth

Behavior of Captive Wild Animals

Hal Markowitz
Victor J. Stevens
editors

Nelson-Hall nh Chicago

Library of Congress Cataloging in Publication Data

Main entry under title:

Behavior of captive wild animals.

Bibliography: p.
Includes index.
1. Animals, Captive—Behavior. 2. Predation
(Biology) I. Markowitz, Hal. II. Stevens, Victor J.
III. Brett, Linda P.
QL751.B364   599'.05'3   77-18156
ISBN 0-88229-385-0

Manufactured in the United States of America

10   9   8   7   6   5   4   3   2   1

Contents

Preface

The authors of this book agreed that all chapters would represent previously unpublished materials. This decision stimulated two worthwhile outcomes. The organization of materials according to a new thematic unity lends cohesion to each section; and, perhaps most importantly, the authors could uniquely express and convey their distinctive enthusiasms about their own major areas of endeavor.

The chapter by Schusterman and the one by Fentress, Field, and Parr both place significant emphasis on communication. Clearly, however, their orientations differ, and not only in the animal order chosen for study. One gains from these sections a real appreciation for the authors' conceptual framework. And it is precisely this sense of "flavor" emanating from each laboratory's research that forms, we believe, an important contribution of this text.

Three of the chapters describe research accomplished largely in zoo settings. The applied free-lance work of Bill Myers in designing behavioral exhibits for several institutions exemplifies the power of operant technology in eliciting behaviors seldom seen by visitors to most zoos. From a different perspective, Markowitz and Woodworth emphasize that the use of similar techniques may give captive animals more control over their own environments. Stevens' work demonstrates the applicability of laboratory techniques to much less isolated zoo environments, and most readers will be startled to see how effectively "laws of learning" can be studied in a noisy, institutional atmosphere. Certainly, Stevens' chapter will give pause to those who tend to discard any data obtained outside the pristine laboratory, and will solace those who must work in complex applied milieus.

Questions about cognition in nonhuman primates have often been explored in superficial or highly anthropomorphic fashion. Essock and Rumbaugh demonstrate that such need not be the case in a chapter which combines a hard-nosed, data-based approach with interpretations carefully related to that data. Their freshly organized work lends renewed impetus to research efforts on the comparative measurement of cognitive capabilities.

The two chapters dealing with predatory behavior offer carefully elaborated experimental studies *and* some practical approaches to husbandry as well. These contributions by Gustavson, Brett, Garcia, and Kelly and by Cheney share compassion for the species being studied plus rational attention directed toward problems of control in the contemporary environmental context.

If one characteristic links together each contributor to this volume, it is this last-mentioned set of attributes. One cannot help but feel the researchers' love for their animal subjects as well as for the data derived therefrom. This is a book by people who like what they are doing.

Hal Markowitz, Ph.D., Director,
Oregon Zoological Research Center

Predator-Prey Interactions

Carl D. Cheney*

INTRODUCTION

Predators discussed in this chapter are defined as animals that kill other animals for food. My major points are that predation is an omnipresent bio-behavioral phenomenon; that predation has not received the research effort it warrants; and that many captive predators would be more healthy, visible, and attractive if they were used to gather data on predator and live-prey interaction. In treating these issues, some predator-prey experiments conducted at Utah State University using birds of prey, such as owls and hawks, and carnivorous mammals, such as foxes and coyotes, will be described. We believe that these studies suggest some general principles of interspecies interaction and that many of them may serve as prototype experiment-display examples.

Killing and eating other animals is, of course, no more offensive to the predator than is eating grass to the herbivore or than seed-eating to the granivore—nor should it be offensive to the observer in this context. Although predators do much scavenging, their survival in the wild depends mainly upon their capture and eating of live prey. The attempt to understand the processes by which predators live and utilize their food resources is not only

*Carl D. Cheney is a member of the Department of Psychology and Institute of Animal Behavior at Utah State University, Logan, Utah.

justifiable in its own right as a biological question, but such research may also provide insights into how man, the ultimate predator, might better operate to conserve his own limited resources. In addition, many concomitant benefits can be gained from captive predator-prey studies, such as viewer education and increased predator activity and health.

Research studies of predator and live-prey interaction are infrequent and little publicized—primarily, I suppose from reluctance to purposely expose a relatively defenseless prey animal to almost certain death and then to write detailed results objectively. Exceptions exist, of course, but generally such work has concerned limited aspects of predation (e.g., Fox 1969; Eisenberg and Leyhausen 1972). Economics may also help account for the rarity of experimental studies using live prey. It is more expensive to maintain live-prey stock than to provide prepared chow, even though total predator subsistence on live animals is unnecessary for research purposes. I feel, however, that an absolute necessity exists for conducting such experiments if we are to obtain accurate data and insights essential for an understanding of predation. Therefore, given the presence of many avian and mammalian predators in zoos and other displays, I submit that behavior with regard to hunting, detecting, stalking, pursuing, and capturing live-prey animals both should and can be easily studied, and that the results will be of more than academic interest. Of the four basic components of predation described by Holling (1959), the two associated with individual predator and prey characteristics are of most interest here. The other two components dealing with population dynamics are primarily relevant to free-ranging animal research only.

The study of interspecies behavioral phenomena, as with most scientific problems, lends itself to the use of a model system approach. A model system in this context refers to a relatively simplistic setting with easily managed environment, predator animals, and prey animals, e.g., an owl and mouse or a fox and chick, as opposed to a wolf and moose or a lion and warthog. An excellent example of an extremely basic model system is that of Slobodkin (1968), who used populations of flatworms, *Daphnia* and *Hydra*, in a series of studies, results of which allowed for development of a mathematically optimal strategy for being a good predator. The use of fish subjects in predator-prey studies (e.g., Beukema, 1968) has received little attention; fish may, in fact, be very good subjects with which to initiate a public display, since these organisms usually engender less viewer emotion. A model predator-prey system may not only provide specific data regarding particular individuals involved, but can also facilitate development of general principles which will be applicable to a wide variety of predator-prey questions (Slobodkin, 1968). My point here is to emphasize that in studying captive predators, if experimental design and methodology are objective, basic, and not forcefully species-specific questions can be investigated by use of very few individuals and/or species; and results will, in fact, have relevance to a wide variety of animal interactions. Using experimental designs wherein potentially critical variables are manipulated as opposed to nonintervening observation, many predator-prey encounters can be compressed into a short time, conditions such as deprivation can be controlled, and attempts at replication are possible.

It is apparent that use of captive individuals as research subjects negates, for the most part, a strictly ethological approach wherein one observes free-ranging organisms and their happenstance interaction in a natural setting. Complete instances of predator-prey encounter in the field are not often observed and are even more rarely recorded and reported in detail. Notable extensive exceptions in the literature include Errington 1967; Schaller 1969; van Lawick and van Lawick-Goodall 1970; and Kruuk 1972. (This is not an exhaustive list, only representative.) Relative lack of field data is partially due to the incredible ineffectiveness of most predators. For example, Errington (1967, p. 45) describes a report of a European Sparrowhawk that was successful in only twenty-two of more than five hundred attempts at predation. The bird, of course, was eating more often than this. But an observer noting a rate of twenty-four attempts per day would spend more than twenty days of observation without recording a hit. Mech (1970) reports the outcome of 120 "detectionstalks" of moose by wolves that terminated in only 5 percent kills (roughly similar to the hawk's record). Predators' ability to tolerate many failures in hunting can

probably be used to advantage in establishing programs of predation display. Also noteworthy is the fact that in the wild, the
predator cannot quit hunting unless it can subsist on carrion.

Along with the paucity of extensive natural history reports,
even fewer manipulative studies of predator-prey interactions are
available in the behavioral literature. Some obvious exceptions in
terms of controlled experiments are primarily avian studies and
include, for example, Royama 1970; Kaufman 1974a,b; Mueller
1974; Ruggerio 1975; and Snyder, Jensen, and Cheney 1975. Given
the current concern of conservationists and others over mismanage-
ment of predators (e.g., Pringle 1975; Olsen 1971; Howard 1974;
Erickson 1972) and the paucity of sound biological information,
researchers might do well to consider this pregnant topic and
concentrate more effort on the ubiquitous phenomenon of predation.

Almost every wild animal at some time in its life is a potential
victim of predation, either when young and helpless or foolish, or
when weakened because of old age, injury, or illness. Few wild
animals die quietly in their sleep. They are either abandoned,
rejected, or pushed into inadequate habitat by their own kind and
thus fall victim to a predator; or, as a result of the normal aging
process, become defenseless. This is true even of the largest predators (Schaller 1969). The point is that predation, as defined here, is
a pervasive process throughout the animal kingdom (Ewer 1973).
This is clearly indicated by reports of predation with such diverse
species as paramecia (Luckinbill 1973) and chimpanzees (van
Lawick-Goodall 1971, p. 197). Predation always enters promi-
nently into any description of wildlife ecology (e.g., Emlen 1973).
This being so, the topic deserves much more empirical attention
than it receives.

Unfortunately, the public view of predation is most often
dramatized via television documentaries or commercial wildlife
films. Such concoctions are often replete with anthropomorphisms
and inaccuracies. Even many purportedly scientific reports are
often extensive reconstructions of circumstantial information
which might include a synthesis of scattered observations, stomach
contents, scats, pellets, and tracks. It is mainly from these kinds of
sources that the public generates attitudes and understanding
regarding predator-prey interaction. Furthermore, researcher bias

may also result. For example, stomach contents rarely differentiate predation from scavenging. Few people have witnessed a complete predation episode.[1] Thus arise myths and legends which tell of bloodthirsty, wanton killers. A contrary indication of the lack of wanton killing is the report that coyotes penned with sheep have gone as long as thirteen days without food before killing.[2] Furthermore, from such extravagant stories may come man's apparent reinforcement for hunting, trapping, and penning powerful and potentially dangerous animals. It is also largely on the basis of unfounded, biased, and generally inaccurate reports that predators get labeled as pests which supposedly compete for man's resources. Misconceptions about damage and threat to wildlife and human property by predators are widespread (e.g., Olsen 1971; Burk 1973; Pringle 1975). If predation is to receive the objective critique and rational consideration required for thoughtful political, ecological, and economic management, a concentrated effort of education and research is needed. Sound data from carefully conceived and conducted experiments are required to provide input to state and federal wildlife-managing agencies (Wagner 1972; Hornocker 1972).

Education of the public, especially young people, is essential in order to explain that predation is the only way of life possible for some species. Development of a wildlife ethic is most easily accomplished with children, who have little vested interest in either total destruction or uncritical protection of animal individuals and species. It is important for a complete understanding of predation to realize that predator activities serve a positive, essential function in nature. A significant point is that man's intervention, in the form of prey-habitat perturbation, often unbalances the delicate predator-prey ecosystem, thus wreaking havoc and frequently forcing predators to prey upon domestic animals. Predators, existing at the top of the food chain, are very susceptible to ecological disruption. We are at a point in history where we can continue eliminating predators, and suffer the consequences of our actions, or we can exercise foresight and initiate some drastic modifications in our own behavior.

1. This statement does not, of course, apply to hunting, fishing, or slaughterhouse activity where man is the predator.
2. P. Lehner and F. Knowlton 1975: personal communication.

Hardin (1974) provides an analysis of what human population management, or lack of it, will mean to the survival of man. The conclusions reached are not far removed from what "naturally" happens in a free-ranging predator-prey system. The predator looks out for itself by limiting its own population and by utilizing alternative food resources. In terms of management at the population level, Howard (1974) suggests that in lower species of prey animals, presence of a predator actually leads to an increase in prey numbers. Hornocker (1970) demonstrated that given a stable cougar population, the number of mule deer and elk, which are the cougar's primary prey species, actually increased. He suggested that this apparent contradiction was possible because the presence of cougars kept the prey animals more active and evenly distributed, thus reducing local overbrowsing of vegetation and thus reducing starvation-related deaths. This point is important to emphasize to those who would eliminate predators of game species. Predators are not, in reality, prey-population control vectors. They do not eat themselves out of house and home (Snyder and Cheney 1975b). The point of view presently gaining acceptance is that predators operate almost exclusively on the surplus members of a prey population and in reality provide a beneficial service for most prey species (Mech 1974). The combined factors of habitat and intraspecies sociology are by far the most important variables controlling any prey population. If predation has any serious effects on populations, such effects probably function to dampen the natural oscillations that occur in most prey populations (Errington 1967) and to increase local species diversity (Parrish and Saila 1970).

DATA IN PREDATOR-PREY STUDIES

When embarking upon a study of predator-prey interaction, it is important to know the social structure of the prey population and the effects of such structure on the vulnerability of individual prey animals. This sort of information generally comes from field studies, but can suggest problems for laboratory study. Errington (1967), for example, describes the intolerant nature of territorial muskrats and how this behavior forces many individuals to attempt existence beyond the threshold of environmental security. By virtue of agonistic pressure from their own kind, these excluded individuals have a low life expectancy with or without the presence of their major predator, the mink. Therefore, one might conclude that potential prey animals unable to control a territory within preferred habitat may be forced into activities leading to their selection by predators. This conclusion suggests that cover, access to food, prey density, extensive movement, etc., can be important prey variables for empirical manipulation. Rudnai (1973, p. 85) describes how zebra—apparently because of their requirement for standing in water in order to drink—often frequent a riverine habitat. Vegetation in such areas disallows adequate cover for hunting lions; thus, because of this evolved peculiarity, the zebra are virtually immune from lion attack at this time. Such field-based reports are useful in suggesting the types of independent variables that are critical for manipulating various species in a controlled study.

Metzgar (1967) demonstrated that individuals selected first by predators from a rodent population were transient animals without a territory and were therefore more active, conspicuous, and hence vulnerable. Mueller (1974) argues that the "odd" individuals within a population of rodents are more susceptible to selection than the "normal" phenotypes. These specific prey characteristics are obvious potential independent variables. However, if this idea is pursued, one would rarely, if ever, expect to observe morphological mutations because they would have such a short life expectancy. Within this context, Coppinger (1970) indicates that novel prey items are actively avoided when first encountered. Such avoidance would be expected from the food neophobia findings of Garcia, Hankins, and Rusiniak (1974) and others. If neophobic reactions are present it is obvious how changes in prey morphology might become established at some level in the population. It appears that there are characteristics impressed upon an individual from its evolution as well as from its membership in a species society that contribute to its vulnerability.

Other prey characteristics for researcher manipulation which might influence probable initiation or outcome of an individual predation incident include presence of an acute injury leading to abnormal movement, a genetic mutation causing modified coloration, or small size due to youth or other circumstances. Kruuk (1972, p. 154) describes a case where the horns of some wildebeast were painted white; in the following months, almost all of these individuals were specifically selected and killed by hyenas. Such extraordinary appearances may or may not benefit the prey animal depending upon the particular predator and prey items involved. Snyder (1975) found that movement was critical in prey selection when large items were offered. The question requires a great deal more empirical study.

Experience is an important variable in terms of how a potential prey item might respond to attack and its subsequent resistance to predation. For example, given success at defense in the past, a prey animal may assume a more aggressive attitude in future encounters. Mech (1970) mentions this aspect in regard to moose and wolves. If a moose successfully repels a wolf attack, the next time wolves approach, the moose will orient or even move toward the pack; generally the wolves will then detour and move on. On the other hand, some experience with predators may be debilitating. Hamilton (1974) has shown that psychological stress of visual exposure to a cat predator can lower the resistance of mice to parasites. A captive situation behooves the researcher to avoid exposing prey to predators prior to the actual trial. Repeated presentation of nonselected prey is also ill-advised. Coppinger (1970) has shown that predator experience, or lack of it, with a particular prey item can lead to either attack or active flight. We have found the number of trials in which animal protagonists are exposed to one another greatly influences the actions and reactions of both.

Incidents in which the prey behaves assertively toward the predator may be expressive of the evolution of distraction displays. Many bird species show bizarre activities when their nests or young are approached; and the fact that these displays exist and are apparently functional also reinforces the idea that abnormally moving prey are unusually attractive to a predator. By suggesting which prey-associated characteristics are critical in eliciting predator attack, the use of a prey species that engages in distraction display would be very instructive in an experimental analysis.

Another source of data in predator-prey interaction research involves the actions, characteristics, and experiential backround of the predator. By using a uniform prey item and standardizing other features of the environment, observers may compare predators of different ages, species, experiential histories, sex, etc., in terms of their hunting strategies and efficiency. An extensive list of independent variables can be generated for experimental manipulation so as to determine the effects of each on the dependent variables of predation strategy and hunting and capturing methods and success. For example: litter size and sex ratio, training techniques modeled in hunting by the parents, dietary intake during youth and subsequent prey preference, social structure of a confined pack, deprivation, and prey pre-exposure may each play an important, distinctive role in predator-prey interactions. In the young of some predator species—for example, cheetah (Eaton 1970) and bear (Luque 1975)—observing the mother hunt and kill is essential for development of survival behaviors. In addition, strategies used in prey selection by experimentally naive individuals (i.e., wild-trapped adults) may reflect the evolutionary influence of trophic levels for the entire species. In regard to individual predator characteristics, recent experiments have investigated the effects of prey availability on hunting distribution, i.e., proclivity to alternate resources (Snyder 1974a), proximity of prey to predator (Cheney and Loether 1975a), familiarity of prey to predator (Ruggerio 1975), and pack dominance structure (Walls 1975).

In the study of predation phenomena, it is clear that data can come from either prey or predator alone; for ultimate analysis, however, there must be a confrontation. One cannot hope to analyze nature functionally by studying individuals in isolation, but must eventually consider the association among individuals and populations. The behavior of both individuals in a predation episode is in great measure directly related to the characteristics of each. How one animal acts or reacts in such a situation depends on how its opponent behaves (Eisenberg and Leyhausen 1972; Fox 1969). An example of how data from such an interaction may be immediately and economically useful involves the management of coyote predation on domestic livestock. In dealing with sheep-killing coyotes, one proposed method among many (Linhart 1974) that will also reduce the danger to non-target species and individuals is to place a cyanide collar on selected individual sheep. Coyotes kill sheep, in almost all cases, by a concentrated attack at the throat. The theory is that a puncture in the cyanide collar will release the toxin and kill the offender in the act. The question of interest from a research standpoint is, which sheep should wear the collars? Economics prohibits fitting all animals, so only those individuals most vulnerable or most likely to be victimized need collaring. But how can a husbandman know if there are some individuals whose physical characteristics or behavioral attitudes denote vulnerability to a potential attacker? Even if he could know this, would all attackers respond in a similar way to a given combination of prey-associated variables? These are the kinds of questions, I feel, that require studies of live predator-prey interaction; and results will provide theoretical and immediately applicable data.

I contend that principles which functionally relate predator and prey will have relevance to an understanding of other types of animal social interactions, including both intraspecific and interspecific behaviors (Denenberg, Posche, and Zarrow 1968), and, in addition, will contribute to an analysis of resource use by animals. As an example, consider the factors responsible for grazing patterns in large ungulates, a type of information important to domestic stockmen and managers of wildlife resources. These patterns may, in fact, be based on the animals' evolution with respect to predator defense mechanisms. How such development comes about may become known through research efforts, but exactly how anything might be done to alter it is another problem. Simple possession of such data, however, may contribute in a positive way to efficient land-use management.

Mimicry of a non-preferred prey is one instance in which a defense mechanism has altered feeding patterns of predatory tanagers (Brower, Cook, and Croze 1965). Flock-feeding as an instance of social organization is another example of an apparently important anti-predator device; as a secondary consequence, the flock structure operates as a population control mechanism (Murton 1967).

Social interaction among higher primates, including humans, can often be described in terms of a predator-prey analysis. Some individuals are "born losers," the perpetual victims in our society—the "sucker," "easy mark," "fall guy," butt of every joke.

Determining how some individuals become winners and some losers can very well be treated in a classical predator-prey experimental analysis (Cheney 1974). The problem is not unrelated to the one mentioned above: which sheep are more vulnerable and why?

RESEARCH AND ZOOLOGICAL DISPLAYS

Most captive predators exist in zoos and other displays for public viewing. I feel that several services can be performed by utilizing such captive animals in live-prey hunting, selecting, and capturing experiments. Other researchers have also recommended the use of captive species in various working displays (Markowitz 1975; Chasan 1974). One important objective of a zoological display is to make an animal visible in surroundings that simulate its natural habitat. Such constructions are ideal for the study of predator-prey interaction. Many predators living in a traditional zoo either sleep most of the day or engage in some stereotypical activity such as pacing. Viewers rarely observe these animals engaging in anything remotely resembling a natural activity. By employing procedures during experiments that activate these animals and make them visible to daily visitors, a major desire of most park personnel could be fulfilled. For example, Markowitz (1974) describes a serval exhibit at the Washington Park Zoo; a target is moved about at various angles and heights so that when it is operated by a jumping serval, food is delivered. This display makes these cats not only more visible and active, but also much more attractive and realistic to the observer.

In addition to inactivity resulting from lack of suitable stimuli, there is the feeding regimen. Zoo animals are generally fed once a day with prepared food, raw meat, or "freshly killed mice" usually thrown on the floor. Animals on this behaviorally non-contingent feeding regimen often become fat, listless, and otherwise abnormal; and, in addition, are usually difficult to breed. As an example of a first step to normalize behavior, simply not killing the mice before feeding time could produce a much more dynamic and vital predator display. A clear instance of where live mice could profitably be used, if only for purposes of feeding and activation of an animal, is with owls. No visitor to a zoo ever sees an owl do more than blink or rotate its head. Yet this bird is a magnificently designed flyer and hunter that might as well be stuffed for all the animation it shows in a typical display. Owls easily learn to eat during daylight hours, and larger species will often take four or five mice every twelve hours, day after day. An announced program of such predator-prey interaction times, I venture, would attract many viewers, particularly if some scientific data were to be gathered on each trial and visitors were completely informed of the objectives and variables under investigation. Another project which we have in progress is to allow an owl-mouse combination to free-run. Mice of various colors and both sexes are provided a system of short tubes in which to hide and through which they must travel for food and water. Data of interest include the selection ratio with regard to color, size, and sex, as well as population growth and predator behavior. For example, will light-colored (odd? conspicuous?) individuals be taken first, as indicated by Kaufman (1973)? Will the population increase or crash? What patterns, if any, of hunting will evolve?

Establishing a program of predator-prey research in a public situation would lead to accumulation of scientific data functionally relating the species involved. At the same time, this program would provide the additional advantage of partially overcoming the nonvisibility-inactivity problems of display. A program which required (or, from the animals' point of view, allowed) an animal to earn some of its daily food by actively searching out, pursuing, capturing, and killing live prey would provide a very realistic educational experience for the viewer (experimenter) and be of definite benefit to the predator. A coyote allowed to chase and capture its food would be provided an outlet for unchanneled, natural hunting behavior, which in the zoo is generally manifested as pacing. Such a program would also allow meaningful activity and therefore provide views of an often-graceful animal doing what it is best equipped to do. Traditionally, as another example, the viewer sees a fox as a rolled-up ball of fur resembling a coat collar. The beauty of this little creature, with ears erect and face alert as it zeros in and pounces on a moving target in natural cover, is indeed unusual. Such a sight is nearly as inspiring as the view of a hawk or eagle swooping gracefully from a high perch to strike an unwary or quickly darting rodent. Observing such incidents makes one aware of the grace of movement, richness of color, and the psychological complexities of predator-prey interaction. Such observations might also allow the visitor to identify more easily with the predator for a change, instead of with the prey. Hemingway (1932, p. 5) separated viewers of a bullfight into those who identified with the bull and those who identified with the man. This psychological alignment is of particular political importance when one is dealing with predator-management activities and organizations.

Markowitz (1975) describes a procedure the Washington Park Zoo is establishing in which seals, carrying physiological transducing and transmitting instrumentation, will be placed in view of their natural predator, the polar bear. I don't suggest that these two species be mingled, but that both would be worth watching at the same time. This design would provide data on such overt and physiological responses as orientation, arousal, and habituation in both animals. We have conducted a similar project in terms of monitoring the heart rate of a raptor as it viewed and ultimately selected live-prey items (Cheney and Snyder 1975).

PREDATOR-PREY STUDIES

Several manipulative studies have been conducted at Utah State University with such predators as fox (*Vulpes vulpes* and *Vulpes macrotis*), coyote (*Canis latrans*), raptor (*Buteo jamicensis, Buteo regalis, Falco sparverius*), owl (*Bubo virginianus*), cat (*Felis domestica*) (Ruggiero, Walls, and Cheney 1977), and ferret (*Putorius putorius*) (Snyder 1974b). We would prefer to work with a larger variety of predators and birds of prey, but some species are more easily manipulated than others, and we feel that principles worked out in detail with some model experiments will probably have general applicability in describing interactions of a variety of species. I shall describe some of these experiments and point out how such studies might be employed in a zoo display.

A basic question pervading our work is: "what are the prey-associated characteristics that are functionally related to probability of attack by a predator?" We assume from the outset that such variables do, in fact, exist and that the parameters of these variables are central to individual predator-prey interactions. While behavioral aspects of predation are somewhat complex (Krebs 1973), they lend themselves to analysis just as do most other biological events.

Clearly, not all encounters between predator and potential prey result in predator attack. In most cases, choice is not random, yet simple opportunism could not govern the majority of prey selections. It is also a fact that predators are not so skilled that they can capture prey at will. They often appear to "look over" a potential meal and then continue to stalk it or move on (Kruuk 1972). One question which is the basis for many of our experiments is "what cues were provided by the individual prey that triggered attack or rejection?" It is clearly not true, however, that only prey variables are critical, and such is not implied. Odd-appearing individuals may be taken by the predator under some circumstances (Mueller 1974) or rejected under others (Coppinger 1970; Ruggiero 1975); but what constitutes oddity? Is it nonconformity in a group? Are the prey in question different in color, movement, size, or locality from those generally taken? Kruuk (1972) reported that he could not discern the basis for hyena prey-selection, though he felt it was clear that the animals were making some analysis.

Having tried to reduce the phenomenon to its bare essentials in the hope that eventually order and generality will emerge from the research, we have approached such problems with birds of prey in the following way. Two prey items (usually mice) are presented simultaneously to a perched raptor. Both prey items are equally active and of identical size. One item contrasts with the substrate upon which it is placed; the other does not. Length of time taken by the raptor to respond and which item is selected are the major data collected. Heart rate has also been collected in some cases. Not only do we record latency of strike and which prey animal is selected; we also note the position of both prey items and distances between all animals. It has been noted, for example, that if a really "peculiar" mouse is too close to the preferred mouse, or even comes toward a kestrel that is holding a mouse, the kestrel either will not strike or will actually abandon a kill (Ruggiero 1975). Position of the selected prey with regard to the hawk, being nearer to or farther from, also seems to differentiate between striking strategies of kestrels and ferruginous hawks in our situation. Kestrels tend to take nearer mice; while, all other conditions being equal, ferruginous hawks swoop onto the farther prey. Owls also appear to prefer the nearer prey in our situation.

In attempting to isolate the factors influencing prey selection, a very important feature we have discovered is the extent to which the predator is experienced with the variety of prey choices provided. For example, a recently captured adult hawk will rarely select a white mouse as first choice between a wild phenotype and a white when both are presented on a dark substrate. Apparently the white mouse, being totally new, is not desirable to the hawk and will, in fact, be avoided (Ruggerio 1975). This finding refers to an issue mentioned above concerning selection for "odd" individuals in the population. We would argue that until a raptor has had some minimum number of encounters with a unique but potential prey, it will actively avoid such extreme novelty (Ruggerio 1975; Beukema 1968). Apparently oddity in this sense (or novelty) is initially avoided; but after some amount of experience (perhaps allowing time for development of an "associative searching image"), the strangeness "wears off." The odd item may then be chosen disproportionately often because it may also be more conspicuous (i.e., white mouse on dark substrate). Something like this process may have been observed by Mueller (1974), who ran kestrels for more than eight-hundred trials and found a tendency for selection against oddity. He also reports, as we do (Snyder and Cheney 1975b), an occasional tendency of the hawk to alternate from one prey phenotype to another, as if the bird were seeking a varied diet or, in fact, some novelty. I contend that predator alternation, or switching, insures a sampling of the availability of alternate prey resources (Tullock 1971; Snyder and Cheney 1975b). This behavior may be founded in dietary needs, boredom, or something else; but it does appear commonly across a number of raptor species (Mueller 1974; Snyder and Cheney 1975a).

The hawk and mouse interaction could easily be incorporated into a public display by requiring the hawk to fly to, and land on, a series of three or four perches before a mouse is automatically released. This program would activate the bird and provide the viewer with repeated examples of the bird's flying, striking, and feeding behaviors.

We are also investigating in a highly controlled situation fox and coyote hunting strategies and prey preferences. We use a chamber and procedures designed to isolate various components of

a foraging episode for automatic recording purposes (Cheney and
Snyder 1974). I believe the general procedures would be applicable
to a public display setting because they would activate the predator
and entertain the viewer. Parameters which can be manipulated
include response cost in hunting and capturing, prey species, size
and vulnerability of prey offered, modeling of experienced predator
trials, and probability of detection-capture matching. We employ
three chambers for prey boxes. Each chamber has a visual-access
door, a physical-access door, and a lever for the predator to operate
with its paw (Figure 1.1). Programming and recording equipment
is located in a nearby building, and almost all components are of
local design and construction. Such chambers could easily be
camouflaged to appear as rocks in a realistic display for public
viewing.

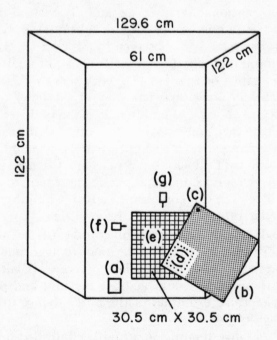

Figure 1.1 Front elevation of prey chamber illustrating operandum location (a)
with visual-access door (b) open and physical-access door (e) closed. (Cheney and
Snyder 1974. Reprinted with permission.)

The general procedure for each single trial is to load all chambers with a prey item, close all doors, then release a predator into the arena. Responding on any chamber operandum, a predetermined number of times (search activity) will cause the visual-access door of that chamber to open. The inside of the chamber is lighted and the predator can then observe the prey item (detection). The predator may then continue to lever-press (pursuit) and eventually open the physical-access door, allowing capture; or, after viewing the prey item, it may move on to another chamber and begin search behavior again. With this chamber and procedure we can vary response cost (hunting effort) by manipulating the number of lever operations required both for initial detection (search) and/or for opening the physical-access (pursuit) door. In addition, requirements can be set such that the visual-access doors on all chambers must be opened before any physical-access door will open. This insures that the predator will observe all three prey items before selecting one. In this way we can contrast such parameters as prey biomass, vulnerability, species, etc. The fact that predators never seem to extinguish hunting-stalking behavior suggests that response costs may be higher than one might expect from a herbivore, for example.

A modification of the general procedure has been to restrain two of the prey items in small cages within their chamber so that pursuit activity could not lead to capture. Sensory stimulation from outside each chamber is equal. One reason for this refinement is to ascertain whether or not the predator will learn that responding, after initial detection of a caged item, is fruitless (since the prey item cannot be captured), and that therefore the predator should move on to another chamber. After running many trials of this type with red foxes, we noticed that upon detection of a nonaccessible prey item, the fox would occasionally continue lever-pressing, open the access door, hop in, investigate the caged prey (often a chick), hop back out, reorient toward the prey, and start pressing the lever again. The sequence could be described as though the fox had associated lever-pressing with obtaining food, and that therefore caged items should somehow become available as a result of continued pressing on the lever. Although an alternate accessible prey was always available in one of the other chambers, a fox was observed to stand looking at a caged item and press forty or fifty additional times. This nonproductive behavior is reminiscent of reports of trained raccoons and pigs engaging in "washing" and "rooting" tokens (conditioned reinforcers) instead of quickly depositing them and receiving food (Breland and Breland 1966).

The procedure of restricting two of the three prey items from capture and consumption allows us to designate a hunting area (chamber) as being 10%, 40%, 80% (or some other percentage figure) likely to lead to success in foraging (Snyder 1974a). Each chamber is thus baited equally, in that smells and sounds are common to each—but differentially, in that one chamber always has a low probability of success, one always has a high chance of paying off, and one is intermediate. In one series, we wished to determine if, in the long run (sixty trials), a fox would distribute its first chamber choices in the form of a probability-matching distribution (Baum 1974), or in a maximizing relationship (Bitterman 1965). This series of studies (Snyder 1974a; Snyder and Cheney 1975b; Cheney and Loether 1975b) revealed that kit foxes are slightly different from red foxes. The red foxes tended to stay with the highest-probability chamber (maximizing), although not exclusively; whereas the kits alternated more. Both species were close to maximizing in terms of making a much higher precentage of choices for the high-probability chamber, but each still sampled the other chambers occasionally. Royama's profitability hypothesis (1970) was supported in this instance; however, since selection was not for the highest-probability chamber exclusively, profitability alone does not account for the results. We feel that a combination of factors is responsible and that adherence to a single concept as the theoretical answer will be difficult to defend.

CONCLUSION AND SUMMARY

Predation is a way of life for carnivorous mammals and birds of prey. They can survive in no other way. Animals under free-ranging conditions live with predation constantly; it is accepted as part of the ecosystem (Emlen 1973). Predators provide input to the factors regulating prey populations (Hornocker 1970), but in most cases the predator contribution tends toward increasing prey population size rather than lowering it (Howard 1974). Knowledge of the principles governing predation will contribute to an understanding of many presently confusing behavioral phenomena.

Many theoretical and applied benefits can result from extensive research with predators, prey, and their interaction.

The major source of predators available for research is within zoological displays. Objectives of the keepers of these animals, and those of researchers needing access to them, are often overlapping. With imagination and careful programming, the public could be brought to accept predation displays as educational, beneficial to the predator, and as sources of biological research data.

Birds of prey show a variety of responses when confronted with hunting opportunities, and the data from such research has led to controversial theoretical positions (e.g., Coppinger 1970; Mueller 1974). It is clear that the use of alternating food-source strategies allow for predator survival in the face of dwindling resources and that most hunting species employ this technique. The Everglade kite is a classic example of an animal's failure to exploit alternate food resources. This raptor preys exclusively on one species of snail. As the swamps are drained, the snails disappear—and so also does the kite.

Examples of using foxes as predators to analyze for prey preferences and hunting strategies, in conditions that are useful as prototypical combinations of experiment and display, show that the concept is feasible and functional.

Data from such studies have suggested that strategies effective in the wild may quickly change in a controlled situation, particularly with predators that are increasing in numbers or at least holding their own. The ability to adapt, improvise, and utilize alternate food sources separates those predators with a future from those without.

2 A Working Model and Experimental Solutions to the Control of Predatory Behavior

Carl R. Gustavson, Linda P. Brett John Garcia, and Daniel J. Kelly*

Predatory behavior is essentially a way of obtaining food and we will treat it as such while taking advantage of a large research literature accumulated over the last two decades. From a broad ecological perspective, control of predation on domestic animals is ideally achieved if we can prevent predation without killing the predator, thus leaving it free to carry out its natural functions in a naturally balanced ecosystem. This is the essence of the method we have been developing. Wagner (1972) presented a history of predation control methods centering mainly on coyote in the sheep ranges; he suggested that these methods have often been ineffective and may have had disastrous results for other species. Any predatory control method must be evaluated within the complex dynamics of the ecosystem, including the one we offer here.

Secondly, we will discuss the factors, particularly palatability, which influence an animal's selection of diet; and we will present evidence indicating that food palatability can be drastically but easily modified. This potential for modification is the basis for our proposed model of predator control. The mechanism for altering palatability simply consists of pairing a specific taste with illness. We propose that the resulting "taste aversion" constitutes the first segment of a two-phase conditioning process which leads to the

*Carl Gustavson is a member of the Department of Psychology at Eastern Washington State College, Cheney, Washington. Linda Brett is in the graduate

predator's rejection of a specific prey animal. Examples of taste-aversion learning come from both laboratory and ethological research and will be discussed in some detail. The second phase of the conditioning process involves the predator's association of the now-repellent taste of the prey animal with its visual, auditory, and olfactory characteristics. The first phase of the process inhibits consumption; the second phase inhibits attack by the predator.

Although food selection is complexly determined, the contributing factors are not equally modifiable in the field situation. In the third section of this chapter, we will examine two types of field manipulation—the use of natural repellents and the conditioning of aversions to prey—which appear to be the most feasible methods of protecting sheep from coyote predation. We contend that conditioning prey aversions is the superior method because fewer technical problems are involved in its use; and also because, unlike most natural repellents, such conditioning is based on a threat to the life of the predator. Such a threat is essential for any control technique to be effective, since predators will ignore nonthreatening noxious properties of a food source when they become sufficiently hungry.

Following this overview of the theoretical basis for our model of predatory behavior, we will discuss our experimental application of the model to the coyote and several other predator species, including hawks, ferrets, mouse-killing rats, and cougars. We will also raise some issues concerning the problems we've encountered in our controlled research with captive wild animals.

KILLING AS A TECHNIQUE FOR CONTROL

Wagner (1972) compared mean annual coyote population indices for the years previous to, and during, the use of 1080 poison laced in animal carcasses. The pre-1080 period represents control efforts from about 1920 to 1950. The use-chronology of each

program in psychology at University of California, Los Angeles. John Garcia is a professor of psychology and psychiatry at UCLA. Dan Kelly is a wildlife researcher from Cheney, Washington. This research was supported by contract under the Washington State Department of Agriculture and research grants from the National Wildlife Federation, the National Audubon Society, the New York Zoological Society, and NIH 1R01 NS 11041-02. We wish to express our appreciation for the help of Mr. Ivan Packard of the Washington State Department of Agriculture and Mr. Michael Thorniley of the Washington Game Department.

control technique during this period varies with the area but includes trapping, shooting, denning, and poisoning with thallium and strychnine in the form of drop baits; and, in the early 1940s, introduction of the "coyote getter," a pipe-shaped device that fired a pistol cartridge of cyanide into the coyote's mouth. The 1080 period extends from about 1950 to 1970 and, while most of the previous techniques were still in use, poisoning with 1080 was the predominant method during this latter period. The population index Wagner used is the number of coyotes taken per man-year of effort, and it is assumed that a decrease in this measure reflects a similar decrease in coyote density. Wagner's data suggests two trends: (1) 1080 was more effective than previous techniques in reducing coyotes; coyote numbers decreased in all areas of 1080 use but one, New Mexico; and (2) 1080 was most effective in reducing coyote numbers in the more northerly states. It should be noted, however, that 1080 poison has been used to a lesser degree in the southern states because Division of Wildlife Services personnel contend that the severity of winter in northern states causes food shortages that force coyotes to utilize the 1080 stations. In the south, however, mild climates and a greater abundance of food may reduce the need for extensive ranging and provide an adequate amount of natural foods (Wagner 1971).

There is reason to believe that the effectiveness of 1080 in reducing coyote numbers would have decreased over time had it not been banned. To quote Wagner (1972),

> It seems likely that the currently used control techniques will decline in effectiveness as time passes. Division trappers claim that the effectiveness of the coyote-getters declined within a decade or so after their introduction, possibly because the animals learned to avoid them. Similarly, 1080 may select an increasingly resistant gene pool much as other biocides have done with other species. Division personnel decry the fact that coyotes too often will not use a 1080 bait, especially if the natural foods are available. The implication of this might be a learning process.
>
> On the whole, the coyote is an extremely adaptable, flexible, and ubiquitous species in western United States. It inhabits a wide variety of environments from the tops of mountain ranges (including winter) to the bottoms of the deserts, and most intervening types. It flourishes on the fringe of agricultural areas, and has moved into the suburban areas of

a number of cities. There is evidence that control has affected its numbers in some areas, primarily the more northern states. But this effect may be less extreme than profound land changes of spreading urbanization and cultivation. There is reason to believe (cf. Clark, 1972; Wagner, 1972; Fredrick Knowlton, unpublished) that food availability and quanitiy may be a more important determinant of density than human control measures in some areas, and in general food availability is probably an important ingredient in coyote numbers in all areas. Clark (1972) and Wagner (1971) concluded that if more food were available in their Utah-Idaho study area, coyotes would be more numerous even in the face of existing control measures.

Wagner's suggestion that 1080 effectiveness has been reduced directly implies that the coyote is learning to avoid the poison stations because sublethal doses of 1080 are being ingested and the subsequent illness establishes an aversion. However, since it is probable that only a few animals survive the poisoning, effectiveness of taste aversion on reducing predation is minimal. Furthermore, the taste-aversion hypothesis is impossible to evaluate from these data, since 1080 was frequently placed in carcasses other than of sheep.

Wagner suggested that despite intensive population control efforts, with population reductions as high as 69 percent, the coyote is not a threatened species. Elimination of the coyote is not, however, the only goal of predator control. Reduction in coyote density has been justified in the past on the assumption that such reduction in numbers is followed by a like reduction in livestock losses to coyote predation. Wagner (1972) described data on changes in sheep losses as follows:

> These results show little, if any, change in the level of sheep lost between the 1940's and the 1080 period. Among the more northerly states in which some reduction in coyote numbers was suggested . . . three showed some reduction in losses (Idaho, Utah and Nevada), two showed increases (Montana, Wyoming), and one showed essentially no change (Colorado). Among the southern states which showed little change in post-1080 coyote numbers, one state showed decrease (Arizona), one no change (New Mexico), and one showed increase (Texas). Grouped together, they suggest little if any change in sheep losses.

Additionally, it should be noted that Montana, which had the greatest reduction in coyote numbers during the 1080 period (69 percent), also had an increase in losses (2 percent during the same period). New Mexico, the only state that showed an increase in coyote numbers (29 percent), during the 1080 period, showed essentially no change in sheep losses. Based on these data, it appears that elimination of the coyote as a species seems unlikely. Moreover, the assumption that reduction of coyote populations results in the like reduction of sheep losses to coyotes is by no means supported. Apparently, manipulation of coyote density to control sheep losses is not a justifiable approach to reduction of livestock losses.

But one might raise a question concerning the effectiveness of eliminating specific offending individuals by trapping or shooting. Wagner (1972) stated that no way exists to evaluate the effectiveness of individual control techniques prior to the use of 1080. First, the chronology of each technique is too variable across areas to make a comparison; and second, no data is available on the amount of predator losses for any period when at least some technique was not in use. This problem is further complicated by food-habit studies that have indicated sheep as a rare item in the coyote diet, making discrimination of feeding patterns difficult. Thus we must ask, do specific individuals feed almost primarily on sheep? Or are sheep a low-level food source for the overall population?

One new form of control technique has been developed by Wagner this year (1975). This technique involves use of a collar containing strychnine worn by the sheep. The strychnine is released upon mutilation, killing both coyote and sheep. While this technique would be specific to offending animals, the question remains as to what proportion of the coyote population is offending. It is very tempting to suggest that such a measure would revert to little more than a poisoning technique not unlike the previous ones, with only the replacement of live bait. But this technique has other problems as well. First, as the number of sheep wearing the collars decreases, so does the probability of attack on a collared sheep; conversely, as the number of sheep wearing the collars increases, so does the danger to sheep and handlers. Secondly, compensation for the first problem could be made by placing the collars on the most probable targets. The most probable target, and the most marketable portion of the flock, is the lamb. Lambs are not very accessible until after docking; and frequently even the herders are not allowed into lambing flocks. Further research on sheep behavior, however, may show that collar placement on probable target sheep can remove particularly troublesome coyotes.

Killing of coyotes as a control technique for reducing stock losses seems limited in value, whether killing is aimed to reduce population density or as a selective technique. Not only has killing failed to reduce predator losses, but Division of Wildlife Resources personnel feel that killing has placed sufficient pressure on the coyote population to cause an increase in pup production, thus counteracting the pressure (Wagner 1972).

MORPHOLOGICAL AND PHYSIOLOGICAL DETERMINANTS

Diet selection determinants range from the most obvious to the extremely subtle, requiring many years of research to differentiate separate organisms. Major ecological distinctions are based on diet selection. All herbivores, carnivores, and omnivores have certain morphological characters associated with each trophic level. Within each group, specialized characteristics have developed that allow feeding on different sources of food within the habitat, e.g., the tree-feeding giraffe and the grass-feeding zebra. Not all differences are so obvious, however, nor are the mechanisms necessarily understood. In a discussion of the use of ecological characters in taxonomy, Mayr (1969) cites several examples of differences in diet selection as being the primary distinction between sibling species. Kohn (1959) demonstrated that all members of the genus *Conus* (a colorful conical mollusk) in the Hawaiian Islands differ ecologically from related species. Two sibling species were both found to feed on nereid polycheates. However, one species (*Ebraeus*) fed almost exclusively on one nereid species, while the second species (*Chaldaeus*) fed, again almost exclusively, on another nereid species. Wagner (1944) demonstrated that two species of fruit fly, *Drosophila mulleri* and *D. aldrichi*, live sympatrically on the decaying fruit of the cactus *Opuntia lindheimeri*. Each species differs significantly, however, in its preference for certain yeasts and bacteria.

Many facets of an animal's diet can be predicted from special-ized characteristics such as size or shape of the animal, size and shape or even arrangement of the teeth, or highly modified loco-motor structures. As in the example of the cactus feeding *Droso-phila*, however, differences are often so subtle as to make the fact of differing diet more noticeable than differences in morphology. Furthermore, closely related species which feed at similar trophic levels may have characters that relate them to entirely different groups. Beidler (1962) measured the relative taste response (i.e., the sum of electrical potential measured at the chorda tympani nerve) to sodium chloride and to potassium chloride. He presented data for two carnivores, the dog and cat; three herbivores, the rabbit, hamster, and guinea pig; and one omnivore, the rat. Beidler concluded that no two species respond alike, but that carnivores respond to sodium chloride and potassium chloride in a fashion opposite to that of rodents. Rabbits, being lagomorphs, not rodents or carnivores, respond still differently. While these differences affirm the evolutionary relationship of these animals, the picture is confusing in terms of trophic feeding level, showing the herbivore rabbit as more closely similar to carnivores than to the omnivorous rat. Beidler's data point to the necessity for examining each species and its habitat for available foods, and the extent to which those foods are exploited by that species.

Food Source Availability

Availability of a food source in the habitat has drastic effects on composition of the diet in most animals. Again, the entire spectrum of food availability and its effects ranges from the ridiculously obvious to highly sensitive subtle mechanisms for adjusting the diet as foods become more-or-less available. For the koala bear, whose diet consists exclusively of eucalyptus leaves, the density of eucalyptus in the area has direct bearing on the extent to which that area is inhabited, and on the amount of area that a specific koala requires in order to feed. Most animals, however, do not depend upon a single food source, but rather adjust their diet to daily, weekly, and seasonal changes in food availability (Marler and Hamilton 1967).

Holling (1959) described increased utilization of European pine sawfly larvae as a food source for both mice and shrews as the

larvae drop from trees and spin cocoons in the dust and needles on
the forest floor. The larvae become available in late June and last
until the adult insects emerge some four months later. In laboratory studies, Holling (1953, 1959) described various aspects of this
feeding situation. Immediately following the placement of cocoons
2 centimeters deep in the sand, *Peromyscus* (deer mice) began
accurate digging at placement locations. A three- to four-day lag
occurred, however, before retrieval of cocoons reached a maximum,
implying that some learning was required for full exploitation of
the resource. Furthermore, Holling found that the mice learned to
distinguish between healthy and infected cocoons, and that under
field conditions the larger female cocoons were more frequently
preyed-upon than the smaller males. He suggested that this was
accomplished through use of olfactory cues. In the laboratory
environment, he investigated the effect of cocoon density on the
consumption of two alternative foods supplied in abundance—one
a preferred food, sunflower seeds, and the other a less-preferred
food, dog biscuits. As density of cocoons increased, consumption of
the alternative food source decreased. A plateau was reached,
however, in which the number of cocoons opened per day did not
increase with increased density. The height of this plateau was
dependent on the type of alternative food present, with the sunflower seeds resulting in a lower plateau than the dog biscuits. At
very low cocoon densities, the cocoons were not exploited, as
though they were not worth the effort; and some critical density
was required to elicit cocoon-seeking with either food alternative.
On the other hand, when the dog biscuits were available, the
density necessary for exploitation was lower than for the sunflower
seeds. Tinbergen (1960) suggested that this shifting of diet with
changes in prey density allows the animal, without excluding
other sources, to keep track of the availability of foods, providing a
rapid mode of change for full exploitation of new food sources as
they arise. He suggested that the great tit has a limited number of
specific "search images" which it can maintain at one time, and he
found that the tit concentrated on a few food items that provided
the greatest yield. This exploitation of highly dense prey never
completely replaced use of low density prey, however, so that
changes could be made as food sources varied.

EFFORT REQUIRED

The ease with which a food can be acquired is obviously
related to the availability of that food source and therefore its
density. That is to say, as a food density increases, less time is
needed to acquire it, and contact becomes more frequent. But this is
not the sole consideration in determining the ease with which a
food source can be exploited. While predator species have evolved
mechanisms to increase the extent to which they can exploit a food
source, prey species have evolved mechanisms to prevent that
exploitation, or at least to minimize predation. These mechanisms
involve cryptic coloration, texture, behavior, and the presence of
noxious stimuli. All such mechanisms are effective in reducing
predation, as exhibited by the extremes of evolutionary design that
some animals display. Many animals have evolved elaborate
combinations of camouflage and behavior which increase difficulty of location by predators. The freezing behavior of a lone fawn
in combination with its spotted coat makes visual location difficult. The treehopper is shaped like a thorn on a rose bush and
places itself on the rose stem to fit the pattern placement of thorns.
Behavior of this insect is apparently maladaptive, however, since
when it is approached, it hops. Many available examples show
mimicry of environmental surroundings, and the frequency of
such mimicry attests to its evolutionary success. The extreme of
this tactic is illustrated by those animals capable of altering
coloration to match their surroundings. In many cases, potential
prey species have evolved protective armors that increase the
difficulty of killing and consuming them. Examples include the
armadillo and the turtle. A third mechanism for increasing the
difficulty of exploitation involves usage of visual, auditory, tactual, olfactory, or noxious gustatory stimuli. In many cases, prey
species possess the ability to spread noxious materials into the
surrounding area or can direct them toward the intruder.

In examining these defensive mechanisms, it is useful to think
of predatory behavior in terms of its appetitive and consummatory
phases. It is clear that visual, auditory, tactual, and olfactory cues
may interfere with the appetitive phase of feeding behavior, while
gustatory stimuli primarily affect consumption.

Smith, Gustavson, and Gregor (1972) and Gustavson (1972) have examined the topography of the pigeon's unconditioned response to shock and the key peck response. They concluded that the pattern of the unconditioned reaction elicited by shock was incompatible with the behavior necessary to produce a key peck response; and that, in the presence of shock, the pigeon was incapable of producing the key peck. In this situation, the key peck response is an appetitive behavior that is prevented from occurring by the defensive reaction to the presence of a noxious stimulus. Many prey species in nature have evolved mechanisms that present just such noxious stimuli which elicit escape behavior incompatible and interfering with continuation of the appetitive approach behavior of the predator. The owlshead butterfly has spots on its wing that resemble the eyes of an owl. When approached, the wings are expanded, displaying the image of owl eyes and eliciting escape, rather than approach, from the predator. The electric eel generates a field of electric current in the water around itself, resulting in nothing less than a shocking experience for the predator. The skunk presents a strong olfactory repellent.

As suggested earlier, most predators maintain a varied diet; and while the major portion of the diet shifts with food availability and ease of capture, the process of selection *never results in the exclusion of other species as food sources.* De Ruiter (1952, 1956), Prop (1960), and Tinbergen (1960) have pointed out that during food shortages, even unpalatable species become a staple diet. In this instance, the conspicuous coloring of unpalatable insects becomes disadvantageous to them (Marler and Hamilton 1967).

PALATABILITY

The ease with which a prey species can be captured within a habitat is related to the extent of its own capacity to interfere with the appetitive phase of the predator's feeding behavior—either by making itself difficult to observe through camouflage combined with wide dispersal of its members, or by its capability of eliciting escape behavior from its predators. Visual, auditory, tactual, and olfactory stimuli of prey can interfere with the appetitive phase of feeding to the extent that the predator can perceive these stimuli at a distance. Gustatory stimuli, however, require the predator to have the prey in its mouth in order to perceive the stimulus; in this case, the appetitive phase has ended and the consummatory phase has begun. For this reason, palatability of a species cannot be considered in the same context of diet selection that deals with ease of exploitation. In fact, biologists have believed for years that unpalatable species use mechanisms of early-warning coloration that increase their conspicuousness, thus forewarning predators that they are not a tasty morsel. More than a century ago, Alfred Russel Wallace described caterpillars of the South American sphinx moth, *Pseudosphinx tetrio*, as gaudily colored in the extreme. Upon approach of an observer, both head and tail of the caterpillar lash to-and-fro. These caterpillars, unlike other insect larvae, are gregarious and can be found lashing simultaneously while covering entire trees (Brower 1969). Also, unlike the palatable species, the unpalatable species depends on early-warning signals and the predator's previous experience with other members of its species.

LABORATORY STUDIES

The fact that certain foods are preferred to others has been studied in the laboratory since the time of Pavlov and has also been used as a tool for comparative analysis. An example is the study by Holling (1953) on the sawfly larvae described earlier. Many prey species contain or excrete chemicals, such as the glycosides contained in the skin of many toads, that inhibit the consummatory phase of feeding. Laboratory investigation of factors that control palatability, however, indicates that food preferences can be drastically altered (Garcia, Kimmeldorf, and Koelling 1955).

Richter (1943) suggested that animals regulate their diet on the basis of specific needs. Many experiments have demonstrated that if an animal is deprived of a specific vitamin such as thiamine, foods enriched with that vitamin will be selected over foods that are low in vitamin content (Richter 1938, 1943; Garcia, Ervin, Yorke, and Koelling 1967; Rozin 1968, 1969; Harris et al. 1933). This preference for vitamin-enriched food is also accompanied by a decrease in palatability of the old deficient diet (Rozin 1967). Rozin (1969) suggested that in order to account for this learning, one must assume that learning can occur with extended delays. Just such long-delay learning was demonstrated by Garcia, Kimmeldorf, and Koelling (1955), who found that illness followed initial taste by as much as 4 hours. The alteration of food palatability through pairing of taste and illness has been studied extensively by Garcia and associates, and many parameters and neurological mechanisms underlying such modification are being disclosed. In general, the taste-illness paradigm consists of presenting the subject with a distinctive flavor followed by an illness induced by x-ray exposure or by injection or consumption of some emetic drug. When the subject next encounters that flavor, the food is rejected as though it tastes bad or is thoroughly unpalatable. The original work by Garcia used ionizing radiation as the illness reinforcer (UCS); subsequent work by Garcia and his associates, however, has indicated that emetic injections such as cyclophosphamide, apomorphine, and lithium chloride effectively function as reinforcers in the taste-aversion paradigm (Revusky and Garcia 1970; Garcia, Ervin, and Koelling 1966). Furthermore, taste aversions induced by oral consumption of lithium chloride generalize completely to an equimolar (.12 moles per liter) solution of sodium chloride; in other words, lithium chloride has no distinctive flavor that is not already present in meat (Garcia and Koelling 1967; Nachman 1963).

Garcia et al. (1972) compared the effectiveness of the taste-illness paradigm to the more classical stimulus-pain (shock) procedure. Their results indicated that when taste is paired with shock, taste functions like other types of conditioned stimuli (e.g., light flashes, tones, odors). Rats will learn to avoid the taste when the interval between taste presentation and shock is three seconds or less; but when the interval is expanded to thirty-three seconds, conditioning does not occur. Further, the taste signal itself does not appear to become aversive when it signals shock, since the rat will accept the same taste presented in a different spatial context (Garcia, Kovner, and Green 1970; Garcia et al. 1972). When taste is paired with illness, however, the specific temporal or spatial properties of the onset of illness do not as drastically influence development of the taste aversion; that is to say, the taste itself appears to become aversive. Rats will reject the taste when it is presented in situations completely removed from the conditioning environment—and the taste aversion develops in one trial (as compared to the many pairings required for taste-pain associations), even when onset of illness follows presentation of the taste by as much as 4 hours. Hankins, Garcia, and Rusiniak (1973) indicated that odor followed by illness functions as a peripheral perceptual sensor similar to sight and sounds in a stimulus-pain procedure. Furthermore, rats that are mucosally anosmic (i.e., with temporary loss of olfactory capacity) are superior or equal in development of taste aversions when compared to normals.

Garcia and Ervin (1968) suggested a neurological hypothesis to account for evidence that taste-illness conditioning differs from noise-pain conditioning. Special neurological mechanisms, differing from the mechanisms generally thought to integrate distally received and cutaneous information, are available to integrate gustatory and visceral information. Both taste afferents and afferents from the viscera and area postrema project directly to the nucleus solitarius. Garcia and Ervin suggested that the nucleus solitarius could act as an "and-gate" for taste and visceral information before relaying this information to higher brain centers. Auditory and cutaneous afferents project to the posterior thalamus via a system of relays. This system, it is suggested, operates as an "and-gate" for integration of auditory and cutaneous information.

Green and Garcia (1971) demonstrated that illness can also be used to enhance the palatability of a flavor. They injected apomorphine to induce a sudden acute illness lasting about an hour, followed by sudden recuperation. They found that if a flavor was presented prior to onset of illness, it became aversive. If the same flavor was presented following onset of illness but prior to recuperation, however, the flavor was enhanced on subsequent tests, as though it were a medicine. One major implication of this work is that hypothesizing separate receptor mechanisms for each vitamin or a separate receptor for each possible nutritive substance is unnecessary. Rather, a single mechanism that can discriminate flavor, and the ability to sense general malaise and recovery from malaise, is all that is necessary to account for alterations in palatability (Garcia, Clark, and Hankins (1973).

ETHOLOGICAL STUDIES

The taste-illness mechanism of altering composition of the diet is functional in the field, not just an artifact of the laboratory. Brower (1969) showed that naive blue jays will attack and consume monarch butterflies. During larval stages, however, some monarchs feed on a form of milkweed that contains toxin (a cardiac-glycoside). These larvae are not affected by the poison but sequester the material into their tissues. If a jay kills and consumes a butterfly that has fed on this milkweed, it subsequently becomes ill. Upon further exposure to monarch butterflies, the jay will not even attack the butterfly, and Brower reported that one jay would become ill and regurgitate at the sight of a monarch. Brower discussed the further ecological significance of this relationship. If the butterfly contains a lethal dose of toxin, one would suspect that large numbers of jays would die. He suggested that the jay's ability to vomit protects the species from a lethal dose. Therefore, both the butterfly and the jay benefit from the relationship. The monarch is protected by the behavior of a resident jay population that will not feed on monarchs, and the population of jays does not suffer a large mortality rate due to the death of individuals that sample the monarch as a food source. Modification in predator diet that results from such a defensive mechanism could have even more far-reaching effects for mammalian species, since diet of the mother is known to influence the diet ultimately selected by offspring.

The experimental work of Galef (1971, 1972) and Capretta and Rawls (1972) has suggested that taste preferences could be established in the offspring of mammals by two mechanisms. First, the diet of the mother imparts the flavor of that diet to the mother's milk. Through the processes of gustatory-visceral conditioning, i.e., the pairing of milk flavor with recovery from hunger symptoms, the offspring come to prefer that flavor much as the vitamin-deficient subjects previously mentioned preferred their enriched diet. Secondly, weaning pups are attracted to the external feeding site by the mother's presence. By feeding at that site, the pup receives nutrition independent of the mother but determined by her selection. Coyotes, unlike rats, feed pups by regurgitation, allowing a third possible mechanism for established food preferences determined by the mother's diet.

SUMMARY

Final composition of an animal's diet is not determined by any single factor, but results from interaction between all of them. How this information is utilized by the animal to determine final composition of the diet, and to what extent each factor is important, remain questions.

Garcia, Clark, and Hankins (1973) have suggested that in order to evaluate the results of laboratory experiments, one must give consideration to four factors: (1) Animals, whenever possible, act on the basis of information extracted from the immediate experimental session, in the context of what has occurred both in previous sessions and between sessions; (2) Coping behavior comes in biologically meaningful "chunks" guided by that information; (3) An understanding of the natural niche in which the species evolved is necessary to explain behavior in the artificial niche created by the experimenter; and (4) Any explanation of behavioral patterns must ultimately be verified in neurological terms.

These factors apply to field as well as to laboratory situations. The extent to which a given food item is present in the diet of a particular animal is a combination of the information on availability of that food, ease of capturing that food, and the nutrients contained in that food which are available following consumption. The animal extracts information from the immediate situation in reference to previous exposures and to what has occurred since the last exposure. Further, the probability that a given string of behavior will be continued is determined by the type of stimuli received during that string of behavior, considering the type of present information available to the animal and its previous experience with that information. Noxious gustatory stimuli will elicit inhibition of consummatory behavior in naive subjects; while previous experience with noxious gustatory stimuli or malaise caused by ingestion of a given food will result in avoidance of that food, thus interfering with the appetitive phase of behavior based on perception of stimuli received prior to onset of the consummatory phase. Avoidance of these stimuli depends on the extent to which they can be perceived and the organism's bias to make such associations (Garcia, Clark, and Hankins 1973). Bolles (1970), Smith, Gustavson, and Gregor (1972), Gustavson (1972), and Grossen and Kelley (1971) have suggested that noxious tactual stimulus (shock) is avoided if the response necessary to avoid that stimulus is compatible with the species-specific defense response to it. Species-specific defense responses must be evaluated in terms of the niche wherein the species has evolved, and in terms of which noxious stimuli have, in the past, pressured the species to evolve behavior patterns suitable for coping with that stimulus.

In general, naturally repellent stimuli are less effectively avoided than are conditioned noxious stimuli, since stimuli that elicit escape behavior do not present threats to the life of the predator. Further, it would not be advantageous to the predator to avoid one food completely because it is more difficult to capture than another food; and, during shortages of one food, overcoming noxious stimuli may be easier than obtaining typical food—the animal will benefit as long as that noxious stimulus does not subvert the food source in terms of its nutritive value. Also, animals having the ability to associate taste with the internal effects of food will avoid the now-unpalatable taste more readily than they will avoid some naturally repellent flavor or other noxious stimulus. If consumption of the naturally repellent substance is rewarded by nutritive effects, the hungry predator will eventually learn to ignore, or habituate to, the repellent properties of the food. The animal would be far less likely to continue consumption of a food previously paired with illness, however, since the life-threatening nature of the situation presents greater pressure to evolve species-specific defense responses.

Several anecdotal pieces of information support this notion. Skunk odor is noxious and effective in eliciting escape behavior from domestic dogs. Dogs, however, are notorious for engaging skunks—and likewise porcupines—over and over again. Further support comes from the observation that during food shortages, even unpalatable species can become a staple portion of the diet (de Ruiter 1956; Prop 1960; Tinbergen 1960). Two examples that appear contradictory to this position should be detailed. Workers developing techniques to control coyotes report that on first contact with certain toads, coyotes will attack and kill, but not consume them; and on subsequent exposure, the toads will not even be attacked. This effect is attributed to the fact that the toad's skin contains glycosides that are so distasteful that the coyote, after one such exposure, will completely avoid the toads. While this is certainly a possible conclusion, it is also conceivable that in attacking and killing the toad, the coyote receives enough of these glycosides to induce an illness conditioning.

In conclusion, many stimulus properties interact in determining the extent to which an animal will accept a given food at a particular time. Some properties tend to inhibit, while others facilitate, food acceptability. Palatability, abundance, and ease of access are important factors influencing this choice. All factors have different effects and relate to different portions of the flow of behavior. Intrinsic properties of the food, however, tell only part of the story. Also to be accounted for is the biological capacity of the predator to overcome or establish barriers to consumption of food in reference to the predator's immediate internal state, its individual past history, and the evolutionary history of its species.

MODIFICATION OF DIET IN THE FIELD

The first section of this chapter suggested that the killing of coyotes has not been an effective technique to control sheep-killing by the coyote. As mentioned previously, all other modes of control fit within the realm of modification of the determinants of diet, as discussed in the second portion of this chapter. Having discussed these factors and given at least some explanation of their effects, the *following sections will discuss how these factors could be used to modify the coyote diet; and the approach that appears most promising toward such modification.*

The first three factors discussed under diet selection were functions of morphology and physiology of the predator and the availability of food. While these factors are modifiable in the laboratory, application to the field situation would require modification of the entire ecosystem, an undesirable tactic. Furthermore, most of the modifications are simply not technically feasible; or would, in and of themselves, remove the necessity of controlling the coyote. Thus appear two alternative modifications of sheep: either to make them more difficult to exploit or to make them less palatable to coyotes. The following discussion is divided into (1) a critique of natural repellents, and (2) a critique of conditioned repellents.

Natural Repellents

A natural repellent is generally defined as any stimulus applied to, or possessed by, the animal that will repel an inexperienced predator. The purpose of repellents is to modify the ease with which the predator can gain access to prey. Many proposed techniques for control of coyotes are of the repellent type, consisting of olfactory, auditory, gustatory, or physical barriers. In our discussion of factors that determine diet selection, it was suggested that presence of a noxious stimulus functioned to elicit escape behavior in the predator. It was further suggested that the extent to which these repellents were avoided depends on the extent to which they present a life threat to the individual or species.

Possibly repellents that successfully elicit escape behavior can be developed. But the technical problems that must be overcome for any repellent to be useful are especially difficult. First, if repellent is to be applied to the wool, it cannot damage it in any way; and if the repellent is olfactory, it must be removable from the wool. Second, lambs are the mostly highly preyed-upon portion of the sheep population, and, as many wool growers have asked, "Who is going to be out there to put this repellent on the lamb when it is born? The mother?" For the repellent to be effective, it must be present on the animal, and this requirement presents some real problems during lambing season. Maternal behavior of the ewe is strongly dependent on a critical period of olfactory and gustatory contact with the lamb immediately following birth (Scott 1967). Most sheep raisers will not even allow herders into the flock during lambing for fear that any interference could cause the ewe to reject her lamb. Therefore, a repellent must not interfere with the olfactory cues involved in maternal behavior; nor can it be applied during lambing, a period lasting from one to two months and the period of heaviest loss due to predators. Thus, several technical problems must be solved in the development of repellents—and even if these problems are solved, repellents will probably be limited in use to nonlambing periods.

The taste-aversion technique for controlling coyotes faces none of these problems or restrictions, however, because it is applied to the predator and utilizes the "natural" qualities of the prey species as repelling stimuli. Furthermore, the period of lambing offers the greatest advantage for developing the taste aversion, since the sheep are gathered into the smallest range area so that treatment can be concentrated. Tactically, it would be most advantageous to concentrate on this latter technique which is applicable to the lambing period; and then either work toward application of that technique on the open range, or develop an additional technique such as a repellent that could likewise be used on open range. Certain advantages could be gained by extending taste aversions to the open range, e.g., establishment of resident-averted coyote populations; but a combined approach certainly seems viable.

CONDITIONED REPELLENTS

The issue of conditioned versus natural repellents is presented here for convenience at an operational level and is not intended as a distinction between field and laboratory-type mechanisms.

An excellent example of a conditioned repellent functioning to protect not just one but two prey species in the field is discussed by Brower (1969). The blue jay's rejection of toxic monarchs, mentioned previously, has been used to explain the close mimicry of the viceroy butterfly to the monarch. Brower, however, suggested that this instance does not exemplify a unidirectional Batesian mimicry function (i.e., the monarch serving to protect the viceroy alone). Rather, the relationship is an example of Müllerian mimicry. He suggested that the viceroy is also mildly distasteful, and that this distastefulness tends to protect both viceroy and monarch. Developed taste aversion to the monarch further reinforces the protection of both species. This ability of an animal to assimilate the poison of a plant into its tissue and then poison its predator is also found in the desert grasshopper, *Poekilocerus bufonius*. It was earlier suggested that the extent to which a predator will avoid a potential food source depends on the extent to which the prey possesses a mechanism that poses a threat to the life of the predator. It is difficult to imagine a stimulus-pain conditioning situation where pain may threaten the life of the coyote without inflicting permanent damage or death. In other words, if shock was used as a noxious stimulus, the level of shock needed to produce effective avoidance would be dangerous. Research on the taste-illness paradigm has shown, however, that certain drugs such as lithium chloride produce illness sufficient to suppress completely the intake of specific flavors. Lithium chloride may be given in effective, nonlethal doses, and any tissue damage resulting from administration of this drug is seen only after long-term chronic use (e.g., fifteen years as a salt substitute for humans).

Killing has not been a successful technique for control of coyote predation on sheep (Wagner 1972), but modification of the factors that control diet selection seems to offer strong possibilities. Modification of morphological or physiological factors of diet selection seems unlikely and undesirable on the basis of gross ecological interference. Modification of the ease with which the coyote can gain access to sheep has several technical problems that must be overcome; and some question exists as to what extent these modifications will actually result in coyote avoidance of sheep. Stimulus-pain conditioning faces the problem of transforming the sheep into a life-threatening stimulus to coyotes without such threat becoming fact to the coyote, the herder, other people, or nontarget wildlife. The taste-aversion technique for controlling coyote predation seems the most likely candidate for an immediately acceptable prevention technique.

Based on this overview, we have developed a model of predatory behavior which has guided our investigation of the use of bait shyness for predation control. We propose that search, attack, and killing comprise the appetitive phase of the predatory sequence. Auditory, visual, and olfactory cues are used to locate and guide attack on target species; eating is the primary constituent of the consummatory phase of predation and is completely guided by gustatory stimuli; consumption of the prey reinforces the entire sequence of predation. Taste-aversion conditioning changes the hedonic value of food flavor from appetitive to aversive. A two-phase conditioning process has been proposed by Garcia, Clarke, and Hankins (1973). In phase one, the food flavor becomes aversive through pairing of the flavor with illness. In phase two, the auditory, visual, and olfactory precursors to consumption become associated with the now-aversive flavor.

Rusiniak et al. (1976) investigated the mouse-killing behavior of laboratory rats as a form of predatory aggression, since rats which kill mice are known to eat them. To briefly summarize this work, rats were tested for mouse-killing behavior in the home cage or in the goalboxes of a T-maze. All mice were white albino laboratory animals of two flavors, plain or sweet-minty. Sweet-minty mice were flavored by dipping the animals in a solution of 2 grams saccharin + 20 milliliters peppermint extract + 980 milliliters tap water. Rats were poisoned for eating plain dead mice, drinking sweet-minty water, or eating sweet-minty mice they had killed. When subjected to post-treatment tests in the home cage, rats would kill but not consume plain mice if illness was paired with eating the plain mice. Rats averted to either sweet-minty water or sweet-minty mice would kill and consume plain mice, and kill but not consume sweet-minty mice, even though both flavors were presented at the same time in the home cage. Rusiniak suggested that killing of the "treated" mice may have been due to irritative and defensive responses inherent in the restrictive home cage testing. When these experiments were replicated in a T-maze, rats clearly avoided "treated" prey; and 33 percent of the "safe" prey choices were accomplished at the choice point in the maze following sniff samples of each arm of the T-maze. These results suggest that killing and eating are independent behaviors which are differentially cued and can have a common motivational source. But they also indicate multiple causes for oral aggression and suggest that bait shyness will efficiently suppress attack only when associated with the food-getting chain of behaviors.

In reviewing the results of six experiments comparing coyote and ferret performance in a bait-shyness learning situation, Gustavson (1974) concluded that both ferrets and coyotes are sensitive to the taste-aversion paradigm, as is the omnivorous rat.

Subsequent tests with ferrets in both home cage and T-maze experiments, however, suggest that alteration of their killing response may be difficult. When consumption of a prey was followed by lithium-induced illness, some reduction in feeding was noted. Changes in killing behavior, however, were not consistent. When altered, such behavior took the form of attacking and killing the prey with the feet, but did not alter the choice of prey. For the rat, the T-maze appeared to provide an environment where intrusion by the prey was minimized, therefore reducing the amount of

killing due to defensive response. But since the ferret's modal
latency to attack was 4 seconds, possibly the ferret was quickly
returning to a place where reinforcement was previously found;
with the door shut behind it, the ferret then attacked with its feet
the now-"poisoned prey" as an intruder. On the other hand, the
ferret is known as an animal that will "overkill" a prey resource.
Like many other mustelids, it will devastate a rodent population
and then move on to a different area. Furthermore, ferrets will
continue to kill mice even when just having finished eating a meal.
It is possible that, with this type of feeding pattern, killing is just not
a very malleable behavior in the ferret.

Coyotes proved to be much simpler to deal with. Gustavson
(1974), Gustavson et al. (1974), Gustavson and Garcia (1974), and
Gustavson and Gustavson (1974) reported that coyotes previously
shown to kill either rabbits or sheep would completely suppress attack following one or two exposures to taste-aversion
conditioning.

One of the more controversial issues concerns the duration of
this effect. Apparently, two coyotes which we averted are again
killing lambs. According to information we have received, at least
one of the animals was fed lamb and so reacquired a taste for lamb.
Both coyotes were induced to kill by enclosing them in a pen with a
lamb. This result is precisely what we would expect, since, as
mentioned earlier, nutritious effects cause animals to "acquire" a
taste for flavors previously rejected. As we have pointed out in
previous papers, "extinction" of the aversion is accelerated if the
coyote is fed the averted food. And, as suggested by Rusiniak (1974),
if alternative foods are restriced or if contact is forced by penning
coyote and lambs together, oral aggression arising from other
motivational sources may be observed. In any case, these coyotes
were trained sheep-killers, and the main function of a field program would be to prevent nonkillers from acquiring the sheep-
killing habit. Once a coyote becomes a confirmed sheep-killer,
perhaps it will be necessary to remove it from the population.

Another misunderstanding concerns our use of lithium
injections, which have obvious advantages in laboratory studies. A
metered dose which cannot be vomited is delivered to the animal,
but injections are not necessary. Our hamburger-lithium studies
(Gustavson 1974; Gustavson et al. 1974) indicate that a single meal

of hamburger containing lithium in an enteric coating produced a strong aversion for hamburger without use of injection. One coyote (Luna) developed strong avoidance of a live lamb after orally ingesting a single lamb meal treated with lithium, and without any injections. Aversion for deer meat was also achieved by oral ingestion alone in a cougar named Fred at the Hogle Zoo in Salt Lake City, Utah.

We have recently conducted experiments with six captive coyotes as preparation for evaluation of taste-aversion conditioning as a control technique in the field. These coyotes were individually caged and presented every forty-eight hours with the oppor-

Figure 2.1 Laboratory-reared coyotes readily attack, kill, and consume small lambs upon first exposure. However, large lambs are not as readily attacked, and feral coyotes trapped as adults require weeks of exposure before killing even rabbits.

tunity to run about seventy-five feet across an open field, where a live chicken and rabbit were available in large 6-foot-square goal boxes. Four habituation and prey-preference trials were run prior to the treatment. Two of the coyotes (Annie and Meg) would not perform in this open arena, so tests were conducted in their home cages. Three animals (Josh, Pacer, and Draba) would immediately run down and attack both prey and take them back to the home cage to be eaten. Girl would not kill or eat chickens, so canned dog food was substituted for live chickens. She would then run down and kill the rabbit, carry it to the canned-food goal box, eat the dog food, and then return with the rabbit to her home cage.

Figure 2.2 In this sequence, the coyote runs to the rabbit's choice side of the T-maze (print 1), kills the rabbit (print 2), and returns to the home cage (print 3) to consume the rabbit. Following consumption of lithium chloride–laced rabbits, the coyote consumes only dog food in the opposite arm of the maze (print 4).

Girl, Pacer, Annie, and Meg were given rabbit carcasses laced with 6 grams of lithium chloride dissolved in 50 cubic centimeters of water. Vomiting occurred within thirty-five minutes after consumption of the carcasses. Josh and Draba were given a "rabbit-bait package" containing 3 grams of lithium chloride in thirty-two #5 gelatin capsules coated with a cellulose acetate compound. The "rabbit-bait package" consisted of 12 ounces Top Choice dog-burger wrapped in a fresh rabbit hide and sewn together with nylon fishing line. Vomiting occurred in about one and one-half hours following consumption of the bait.

These animals were again tested in the choice situation. Girl ran first to take the rabbit as on previous trials. After smelling the rabbit all over, however, she ran to the dog food, ate it, and returned to her cage. Girl repeated this sequence eight more times within twenty-five days without even approaching the rabbit, and with no further treatment.

On the first posttest choice, Pacer attacked and consumed the chicken. On her second posttest choice, she attacked and consumed a rabbit. She was then given another lithium tainted carcass. Chicken was her choice on the third posttest trial. About 0.5 hour after consumption, however, she started vomiting. We investigated our source of chickens and discovered that they were possibly contaminated with a "fly bait." Subsequent tests with Pacer have indicated that she will neither attack nor eat chickens or rabbits. But when released into the arena, she investigates both and returns to her cage where she is fed a diet of dry food. Both Draba and Josh attacked and killed the rabbit on the first posttest choice and then attacked and consumed the chicken, refusing to eat the rabbit. Josh urinated on the dead rabbit carcass. Each coyote was then given a tainted carcass which it consumed, and both subsequently became ill. On the next choice test, both coyotes repeated their previous performance. But subsequent choice tests 4 months after treatment indicated that rabbits were completely rejected for either chickens or commercial dog food.

Annie and Meg were not allowed a choice-test situation after their first treatment with tainted carcasses. Rather, they were given a bait package to determine whether a coyote that has fed on a tainted carcass will subsequently consume a bait. Both coyotes readily ate the bait package and became ill. Again, follow-up choice tests have indicated that the animals will neither attack nor consume rabbits, but now prefer chickens. We propose that an appropriate field program would consist of distributing "sheep-bait packages" and tainted sheep carcasses around lambing grounds and across other sheep-ranging areas prior to lambing or introduction of sheep. Possibly the winter months are most appropriate for this method because cold temperatures would tend to preserve carcasses and baits, and possible food shortage during the cold months may force coyotes to feed on these sources.

Effectiveness of our coyote control programs utilizing baiting techniques is contingent on consumption of the desired bait by the target population. Control measures using this method have experienced varying degrees of past success. Strychnine and 1080 baiting programs were highly effective in reaching both target and nontarget species. However, antifertility programs (Linhart, Brusman, and Balser 1968) enjoyed little success due to extensive consumption of baits by non-target species, primarily rodents and ravens.

Control of coyotes by aversive conditioning depends upon a successful field-baiting technique. We know from previous observation that coyotes readily consume the bait package under laboratory conditions, but the rate of consumption under field conditions is unknown.

In order to estimate bait consumption by free-ranging coyotes under field conditions, a study was initiated on September 30, 1974. The study was conducted on a 4.02 square kilometer area of Turnbull National Wildlife Refuge, Cheney, Washington. Previous studies confirmed a high coyote population in this area.

Eight bait packages were distributed among five stations. The bait package consisted of 454 grams of chunked sheep meat enclosed in sheep wool composing a bait package 15 centimeters by 8 centimeters with a three to four week minimum field expectancy. Three of the baits were enclosed in a protective plastic coating which, assuming that coyotes could be induced to consume this type of package, would preserve baits indefinitely. Each station consisted of a smoothed-dirt area approximately 0.76 square meter, in the center of which was placed the bait package. The stations were located along established trails in the study area. Number of baits consumed and the species involved, determined by tracks left at the site, were recorded.

All unpackaged baits were removed from the sites by coyotes within a nineteen-day period with peak removal occurring between the sixth and eleventh days. Packaged baits were not taken, but once the protective coating was removed they were readily picked up.

Station 1 was the most informative concerning the type of bait that a free-ranging coyote will pick up. Initially, a plastic-wrapped bait and an unpackaged bait were placed at the site. The unpackaged bait was picked up on the sixth day, but the wrapped bait remained undisturbed. Replacement of unpackaged bait on the eighth day resulted in its being taken by a coyote the following day. The wrapped bait, remaining undisturbed, was unpackaged on the ninth day and taken on the tenth day. This was replaced with another wrapped bait which was not taken until it too was unpackaged on the twelfth day. Tracks at the site clearly indicated that coyotes investigated the wrapped baits but in all cases rejected them. Similar results were recorded at the other wrapped-bait station, clearly indicating the unfeasibility of plastic-wrapped baits in the field.

Stations 2 and 3 were scented with a coyote attractant (coyote urine mixture, commonly used by trappers) on the assumption of increasing the probability of bait location by coyotes; but its use did not appear significant in increasing coyote discovery of baits. Baits at these sites, in fact, were not taken until the eleventh and nineteenth days, and were the last to be located. We feel that use of this technique, however, should not be eliminated on the basis of these data. Since previous extensive field studies were conducted in this area, the stations were established with prior knowledge of coyote movements. In a less-familiar area, the use of attractants may very well prove valuable in increasing coyote bait location. Considering this possibility plus the limited number of strategies employed in this study, the techniques warrant additional application.

Of particular interest was the method of bait location by coyotes. At Stations 3, 4, and 5, coyotes were recorded passing within 3 to 6 meters of the station showing no indication of interest in the baits. Possibly the coyotes just missed the baits or, more likely, were not familiar with sheep spoor; thus the baits did not provide a salient cue. Tracks at Station 4 indicated that a coyote walked upwind 10 meters past the bait, then returned downwind to retrieve the bait, suggesting the use of olfactory cues for location.

Remains of one bait were located approximately 12 meters from Station 2. The wool, pulled apart, was scattered over a limited area with no evidence of skin or meat, thus suggesting consumption of these—a method of bait consumption closely resembling that observed in our laboratory coyotes. In one instance after a coyote picked up a bait, tracks were followed approximately 0.25 kilometer down a road before being lost. Failure to locate the bait in this distance and absence of bait-remains at other stations suggested that coyotes probably moved some distance from the stations before consuming baits.

The primary problem experienced by Linhart, Brusman and Balsar (1968) in their coyote-baiting programs was removal of baits by non-target species, with rodents and ravens accounting for 64 percent of such removals. Ravens (*Coruns sp.*) are rare in our study area, but magpies (*Pica pica*) occupy essentially the same niche and are extremely abundant. Numerous rodent species and other potential predators, including striped skunks (*Mephitis mephitis*), weasels (*Mustela frenata*), badgers (*Taxidea taxus*), bobcats (*Lynx rufus*), and several species of raptors also inhabit the area. Of these species, however, only magpies were recorded at the bait stations. Magpies pulled small tufts of wool from the baits, but in no instance was the hide penetrated. On one occasion, magpies turned over a bait. Interestingly, magpie activity peaked during the initial phase of the study and showed marked decline after the seventh day. The lack of non-target-species disturbance suggested that baits were coyote-specific.

Finally, we initiated an application of taste-aversion conditioning to a sheep-ranching operation. A 3,000-acre ranch in southeastern Washington was selected for the study. The ranch is located in an area commonly referred to as the "channeled scablands," characterized by rolling topography interlaced with a network of lava-rock buttes and canyons. Dominant vegetation of the scablands is bunchgrass (*Agropyron sp.*) and bluegrass (*Poa sp.*). In heavily grazed areas, cheatgrass (*Bromus sp.*) has replaced

Figure 2.3 This photograph of the experimental ranch vividly illustrates the lava-rock buttes common to the "channelled scablands."

Figure 2.4 This young female coyote was trapped on the experimental ranch. In her frenzy to escape, she chewed the top off the wooden post in the center of the picture and flung herself over the fence where she finally died.

bunchgrass. Ponderosa pines (*Pinus ponderosa*) are scattered sparsely throughout the area. Sections of the ranch are under cultivation. A modified maritime climate prevails with most precipitation occurring during the cold winter months. Precipitation is minimal during the warm summer days (Hall 1972).

The ranch is bounded on the south by the Palouse River and on the north by Washington State Highway 26. Rock Creek bisects the ranch from north to south and converges with the Palouse River on the south boundary. Rock Creek, swollen with snow runoff from January through May, is generally impassable except by bridge on the highway and probably acts as a limited barrier to coyote movement. The Palouse River along the south boundary serves a similar function. The western sector of the ranch consists primarily of rangeland; sheep or cattle were not present, however, during the period concerned in this study. An outstanding characteristic of this sector is one of the ranch's two large, open, carcass dumps, located in a canyon about 0.4 kilometer west of the creek. This dump appeared to be one of several regular coyote feeding areas.

The eastern part of the ranch was the main area of focus for this study. During the winter, 400 to 1,000 ewes are maintained in fenced bottomland pastures to the east of Rock Creek. In 1974–75, an additional 160 holdover lambs were also wintered in a feedlot adjacent to the lambing sheds along Highway 26 east of Rock Creek. At one time, the ranch was a shed-lambing operation; in recent years, however, range-lambing has been practiced. Pregnant ewes are released onto rangeland east of the winter pastures one to two weeks prior to lambing in mid-March.

On January 27, 1975, the study was initiated with the establishment of twelve bait stations throughout the entire ranch. These initial stations were selected on the basis of presence of fresh coyote tracks in the snow along fencelines, at gates, at carcass dumps, or at sites where carcasses from previous ewe kills or natural losses were present. Baits consisted of 340 grams of Top Choice dog food containing 6 grams of LiCl, lithium chloride, and wrapped in a fresh piece of unsheared hide 25.4 centimeters x 15.2 centimeters stapled in envelope-fashion. All carcasses on the ranch were injected with a solution of 1,154 grams of LiCl per 14 liters of tap water throughout the carcass, so that a coyote feeding anywhere

Figure 2.5 Mike Sweeney, a graduate student, injects this ewe carcass with lithium chloride solution.

on the carcass would consume at least some quantity of lithium chloride. Carcasses either decomposed or consumed to such an extent that injection was impossible were, when feasible, sprayed thoroughly with the LiCl solution. By the end of April, approximately 212 lithium baits or carcasses available for consumption were distributed among the established stations and at other sites on the ranch. Aerial examination of the ranch revealed that our baiting was conspicuous.

These stations were checked every twenty-four to seventy-two hours. A bait found missing was replaced, and the area was examined for tracks and remains by walking in expanding concentric circles from the station. Feeding activity on tainted carcasses was photographed and determined by examination.

Figure 2.7 presents the cumulative number of Top Choice and sheep baits consumed by coyotes as reported at weekly intervals from January 27 to April 30. Top Choice bait consumption was initially quite high, with thirty-six baits being consumed in the first three weeks. Consumption of Top Choice baits then decreased through week eight, when no more baits of this type were consumed.

Figure 2.6 This photograph of a bait station on the experimental ranch shows the lithium chloride-laced carcass with two Top Choice dog food baits. Coyote feeding on the carcass can be seen on the rump with wool scattered on the snow amid the tracks.

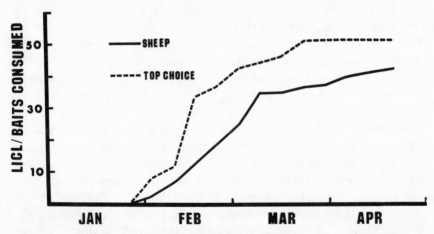

Figure 2.7 Cumulative number of Top Choice and sheep baits consumed by coyotes from January 27, 1975, to April 30, 1975.

Feeding activity on poison-laced sheep, on the other hand, maintained a steady increase through week six. Some of the difference between consumptions of these two differing baits may have been due to initial prevalence of Top Choice baits as compared to lithium chloride-laced sheep; or to inability, in some instances, of determining the number of feedings on a carcass represented by one report. The functions represented, however, are quite similar. Our laboratory experiments with Top Choice/rabbit-hide baits and lithium chloride-laced rabbit carcasses suggested that coyotes will consume Top Choice-type baits once and are not at that time averted to live rabbits. Possibly Top Choice baits are not sufficient to establish aversions in and of themselves since, apparently, two exposures are frequently needed to suppress attack behavior; the Top Choice bait does, however, seem adequate for initial exposure to the treatment, especially when carcasses are not readily or sufficiently available. Furthermore, since coyotes may only consume one of these baits, the number of baits consumed and rate of consumption in a field situation may provide important information on the density and mobility of coyote populations. With this in mind, we suggest that the initial high rate of bait consumption probably represents the resident population of coyotes—including those not only on the ranch, but also transients that frequently traverse the area. Later bait pickups may represent less-frequent visitors or new coyotes in the area. During our previous study on the Turnbull National Wildlife Refuge, magpies were the only other species found to be interested in our baits. This finding was again confirmed.

All coyote sightings were recorded, and notes were kept on the frequency of fresh coyote signs. Initially, single and multiple groups of coyotes were spotted on an almost daily basis, and fresh tracks were common along established trails and fencelines. Coinciding with the drop in bait consumption, however, was a reduced number of coyote sightings until early April, when five coyotes were observed in the early morning with the sheep. Several more sheep-bait consumptions were recorded, and a bitch coyote from a den just over the eastern boundary of the ranch was reportedly shot by the rancher. Coyote activity again appeared minimal. When a trapper came from the Washington Game Department during the last two weeks of April, he experienced difficulty locating recent coyote signs, and only two lambs, a porcupine, and a badger were caught in his traps.

Due to the short time alloted and the lack of an available herd to use as a "sham" control, reliance on past records of the ranch owner was necessary to evaluate changes in rate of predation. This is a tenuous situation at best and greatly restricts the certainty of conclusions. Records had been kept on a monthly basis, however, for at least three years. Figure 2.8 presents the cumulative number of ewes and lambs reported killed by coyotes for the months January through April in the years 1972, 1973, and 1974. During 1972, 700 ewes occupied the range; 600 ewes were present in 1973, and 500 in 1974. We don't believe these records represent the actual number of sheep lost to coyotes, but rather denote a consistent proportion and evaluation based on a personal, but not obviously discernible, set of criteria. First, no effort was made to separate instances where the lamb had been killed or had died prior to being

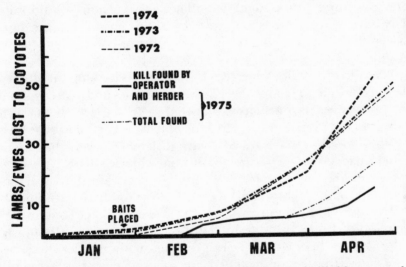

Figure 2.8 Cumulative number of ewes and lambs reported killed by coyotes for the months January through April in 1972, 1973, and 1974; also, cumulative reported kills for January through April 1975 found by ranch operator and herder, and total kills found by all searchers.

fed upon. Furthermore, several instances during our evaluation indicated that different legs from the same lamb were counted as more than one kill. But many cases obviously fit the more usually accepted criteria, such as blood in the wool. Certainly, for our purpose, these differences were not important so long as we could believe that our presence was not affecting the relative judgments of the owner. In past years, all carcasses had been located by the owner and one herder. During our treatment period, however, at least two and sometimes three researchers plus the owner and a herder were locating carcasses. Therefore, ewes and lambs lost to coyotes during our treatment period in 1975 are represented by two lines in Figure 2.2. The solid line represents the carcasses, or pieces of carcass located by either the owner or his herder, that were classified by the owner as a kill. The broken line represents all of the carcasses, or pieces of carcass, found by everyone, thus classified. These lines represent upper and lower limits of the number of sheep classified as kills by the owner; the actual number that would have been found without our presence lies somewhere between. Difference between the losses recorded for 1975 and the mean of the previous three years suggested that taste-aversion conditioning may have reduced losses to coyotes by 54-66 percent, a figure much larger than credited to any previous form of control reported by Wagner (1972).

While our major objective has aimed to develop an efficient field technique to reduce sheep losses to coyotes and save coyotes in the process, we have also investigated the extent to which establishment of taste aversions can serve to control other predatory species. Furthermore, we have tested many different species that might suffer inadvertent exposure to lithium chloride baits. A pair of young timber wolves (*Canis lupus*) were trained to kill and consume both rabbits and sheep. Following this training, the wolves were individually given a sheep-bait package containing 6 grams of lithium chloride in cellulose acetate-coated gelatin capsules. When these wolves were presented with another sheep two days later, the female initiated an attack and was subsequently joined by the male. After bringing the sheep down once and having inflicted one minor wound, both animals rejected the sheep. Two rabbits were then introduced after the wolves had left the sheep alone for six minutes. The rabbits were immediately killed and consumed.

Also, we have researched the alteration of prey preference by aversive conditioning in two North American species of raptorial birds: the red-tailed hawk (*Buteo jamaicensis*) and the American

rough-legged hawk (*Buteo lagopus*). A major part of our interest in
these animals involves their usefulness as models for the behavior
of the golden eagle. Eagles have been greatly maligned by ranch-
men who insist that the birds are a menace to small livestock,
especially domestic lambs, although the actual destructive threat
posed by these animals is almost absurdly small. Golden eagles do,
of course, eat some carrion, and it is likely that their bad reputation
has arisen in part from occasional observations of the birds making
a meal of a calf or sheep either killed by some other predator or
fallen to disease. In any case, our hope is that alteration of prey
preference can serve as a method of controlling whatever predation
problem exists while sparing the lives of the birds. This approach
would certainly be superior to predation control by shooting of the
eagles as illegally practiced in the past.

In addition to its practical potential, conditioned prey
aversion in raptorial birds is interesting on a purely theoretical
level. Assuming that such conditioning is possible in these birds,
what sensory cues are most salient to the raptor in rejecting the
prey? Birds of prey are especially interesting in this context
because, of all the animals one could study, they probably possess
the keenest vision. Moreover, raptors kill prey primarily with their
taloned feet, while mammalian predators kill primarily with their
teeth. Thus, unlike a coyote, the hawk generally has no taste
information to guide its behavior while killing its prey. If taste
proves to be more salient than visual cues in prey rejection, we
might find it more difficult to inhibit killing behavior in birds
than in the coyotes we studied.

The research to date on food-aversion conditioning in birds
indicates that species differences may exist. For at least one avian
species, the bobwhite quail, a visual cue is even more easily
associable with illness than a taste cue (Wilcoxin, Dragoin, and
Dral 1971). Brower's above-mentioned investigations (1969) of
predation on toxic butterflies by blue jays, an omnivorous species,
suggest that both visual and taste information are important in the
birds' rejection of these butterflies. Part of our research activity has
been concerned with determining if strictly raptorial predatory
birds will behave similarly to Brower's blue jays in a situation
roughly analogous to that which leads to the jay's rejection of

monarchs as food (i.e., consumption of the prey animal followed by illness).

The hawks used as subjects in our research were placed in our charge by the California Department of Fish and Game. With the exception of a single male American rough-legged hawk, all were red-tailed hawks. In addition to their behavioral resemblance to golden eagles, red-tails were selected for use in this experiment on the basis of their availability. These hawks represent the most numerous and widespread raptor species in North America. Red-tailed hawks are common in southern California and have a reputation for being "cosmopolitan"; i.e., they can be found closely adjacent to an area as densely populated as Los Angeles. While driving to work on the freeway, one can on occasion spot a red-tail soaring above the nearby hills or over an open field. Although primarily rodent-eaters, red-tails are more flexible in their diet than are most other birds of prey. This adaptability in habitat and food selection undoubtedly contributes to their large numbers. Much of our work to date has been focused on a particularly aggressive and noisy adult female red-tailed hawk, which was found perched in a tree in a local resident's yard. This bird had apparently escaped from whoever had captured her as a young bird.

The male rough-legged hawk in our study represents a rodent-eating raptor species found most commonly in tundra regions, but migrant in winter to the northern and central United States. This bird had been confiscated by the game warden from a person who had illegally captured him, probably in Montana or Wyoming, as an adult. The animal had suffered a broken wing which was never set. He is extremely wary of humans, a trait suggesting the greater part of his life was spent in the wild rather than in captivity.

Before describing the particulars of our research, some important difficulties involved in working with these animals should be mentioned. Many of the research problems we've experienced are similar to those encountered by anyone working with a little-studied animal. Much of one's initial effort must be directed toward learning the "nature of the beast." The scientific literature on bird behavior in general is small compared to that dealing with mammals. It is very sparse indeed in the case of birds of prey and almost entirely anecdotal. Surprisingly, much of the most detailed and useful information we've obtained on raptor behavior came not from scientific reports, but from practitioners of the ancient sport of falconry. This source provided substantial data on raptorial hunting and feeding behavior, which are the main points of our experimental interest, and was invaluable in teaching us how to correctly handle and maintain these animals in captivity.

A particularly difficult matter in studying the conditioning of prey aversions in raptors has been the lack of available information on the digestive system of these animals. Such data can be found rather easily for chickens and pigeons, but these birds are seed-eaters. We have yet to discover data that compare the digestive processes of birds. Time required for a toxic agent placed in the bird's food to pass into the small intestine and be absorbed appears longer by many hours in hawks with a full stomach than in coyotes, and is important in determining viability of the bait-package technique of predator control proposed by Gustavson et al. (1974). A separation of several hours between consumption of the prey and illness might reduce effectiveness of the technique.

The sketchy information available on the particular animals we use in our experiments is another difficulty. The birds have vastly different backgrounds. Some, for example, have been captive from the time they were nestlings, while others have never been in captivity before coming to us. Wild raptors tend to be wary of any interaction with humans, while the more "domesticated" ones will literally eat from the experimenter's hand. Often days must be spent in habituating the wilder birds to eat anything in the presence of human observers. Moreover, in most cases we have little or no information concerning a hawk's dietary history (e.g., what prey animals, if any, has the bird been fed while in captivity?). Since the number of subjects available for our research is, of course, considerably smaller to begin with than the number typically used in laboratory animal experiments, our inability to control for differences in individual hawk histories presents problems in data interpretation. This will become clear when we discuss results of the research to date.

Perhaps the greatest procedural problem encountered in these experiments was determination of what constituted an "effective" dose of the emetic drug; i.e., what amount of dosage will make a hawk fairly sick? Dealing only with this matter required considerable patience and involved several months of work. The fruit of our effort thus far is encouraging, although the greatest part of work on the project is yet to come.

The red-tailed and rough-legged hawks mentioned were averted to bitter-flavored black mice and demonstrated this aversion for more than two months after a single pairing of such a mouse with lithium chloride-induced illness. Both birds have been tested repeatedly since their lithium treatments and have furnished some interesting and rather surprising results. Prior to their illness exposure, these hawks had eaten dead, unflavored white mice and dead black mice treated with quinine. The bitter-flavored mice were eaten more slowly, and consumption was accompanied after the first bites by jerky flipping of the birds' heads from side to side, indicating sensitivity to the taste cue. The bitter-flavored mice, however, were completely devoured on all occasions (within two or three minutes for the red-tail, and within nine to thirteen minutes for the rough-leg).

Following consumption of the second bitter black mouse, illness was induced in each hawk by an intra-peritoneal injection with the toxin. In the red-tail's case, pronounced illness followed injection by a few minutes. The bird vomited several times and looked extremely out of sorts (lethargic, poorly coordinated) during the course of her sickness. The rough-leg, given a much smaller dose of lithium chloride, vomited only once almost an hour after injection. Emesis was the only real indication of illness in this animal. Moreover, the rough-leg ate most of his vomitus a few minutes later, suggesting far less illness than in the red-tail.

The first posttest, the presentation of a dead, bitter, black mouse, followed treatment by ten days for the red-tail and seven days for the rough-leg. Response of the red-tailed hawk was dramatic. The mouse was first thrown on the floor of her cage, as was usual. Before treatment, the bird had pounced upon dead black mice immediately (less than one second), grabbing the animal with her talons in midair or seizing it as soon as it hit the floor. During this trial, she remained on her perch and seemed totally indifferent to the mouse. After several minutes had passed, the mouse was picked up and offered to her from the hand. Usually this bird will take food eagerly when so offered. On this occasion, however, she chittered with alarm and moved briskly to the opposite end of the perch. When pursued with the mouse, she flew around the cage and landed on the floor. The mouse was then placed near the bird's feet. After staring at it for a few seconds, she literally ran away from the mouse, and the trial was ended with the bird never having tasted the animal.

The rough-legged hawk's behavior during his first posttest was similar to the red-tail's. The bitter black mouse was placed next to him on the perch. In his previous experiences with these mice, he, like the red-tail, had grabbed them immediately. On this occasion, however, the bird first stared at the mouse, then ran to the opposite end of the perch, where he remained throughout much of the 15-minute trial. The trial ended without his having tasted the mouse. Immediately following these tests, each hawk was offered a dead, unflavored, white mouse. Each hawk pounced on its white mouse without hesitation and devoured it, demonstrating that the birds were responding to the color cue alone in rejecting the black mouse. Magnitude of the birds' responses suggested that visual cues might be especially important factors concerning development of aversions to poisoned prey in raptorial birds.

The next question addressed was whether the hawks would, like Brower's blue jays, inhibit their killing of prey in response to the visual cue. The birds showed marked individual differences in this respect. The red-tail killed on the two occasions she was offered live black mice. On the first occasion she showed some hesitation in pouncing, but this increased latency disappeared the second time such a mouse was presented. In both cases, however, killing was not followed by consumption of the black mice. The mice had been dipped in bitter solution prior to being thrown into the cage, and the red-tail rejected the mice upon tasting their fur. This tasting consisted of repeated picking up of the mouse with her beak, then flinging it away. Immediately following these tests, the bird killed and vigorously consumed unflavored white mice.

The rough-legged hawk was given four tests with live, bitter, black mice. His behavior during the first test resembled that of the red-tailed hawk. After several minutes of hesitation, the bird pounced upon the mouse and killed it, then rejected it after tasting the fur. When tested the second time, however, the hawk never attacked the mouse. The bird on both of these occasions was offered an unflavored white mouse immediately following end of the trial, and the white mouse was quickly killed and consumed. At this point we were very encouraged—the bird appeared to have formed an association between appearance of the animal and its unpalatability that was sufficient to inhibit predation. This notion was reinforced during the third presentation of a live, bitter, black mouse. Again, there was no attack. The bird then proceeded to "double-cross" us, however, by refusing to attack the live white mouse as well. An unusual amount of uncontrollable noise and activity in the test area had occurred on this day, and the bird was agitated before and during the black-mouse and white-mouse tests; and this excitement may have contributed to the unexpected outcome. We decided to give the hawk some time to calm down and retested him later the same day. This time the bird killed both mice, though it ate only the head of the bitter black animal.

The failure to strongly inhibit killing of the poison-paired animal was disappointing, but it may have been due in part, at least, to the use of subject hawks that had been captive for a considerable period. Unlike wild hawks, which generally do not kill unless hungry, captive ones will sometimes continue to kill even when their crops are stuffed with food. Our red-tailed hawk, for example, once gorged on mice and was then offered additional live mice in her cage. She killed them immediately, but left them lying uneaten on the cage floor. It is possible that the issue of killing inhibition can only be finally settled in a field test situation with truly wild raptors.

The live-mouse tests indicated that both visual cues and taste cues are significant to hawks in rejecting a prey object, and the tests raised questions about possible generalization of the conditioned aversion. How would the hawks now respond to mice that looked different from previously-rejected prey but tasted the same, or to mice that looked the same but tasted different?

In order to separate taste and visual cues, we first presented the birds with a dead bitter-flavored white mouse (different color, same taste) followed by an unflavored white mouse, as usual. Both hawks demonstrated a generalization of aversion on the basis of taste. The red-tail picked at her bitter mouse for almost twenty minutes, finally abandoning it half-eaten. The rough-legged hawk rejected his bitter mouse upon first taste but quickly seized and ate the unflavored mouse. The red-tail, on the other hand, did not pounce on the unflavored white mouse immediately, according to her usual fashion. Instead, she remained "stoically" on her perch, giving the mouse an occasional disinterested glance. Finally, after almost 10 minutes, she descended on the animal and pecked at it repeatedly before beginning to eat. Once she began, eating was vigorous. Here again, both taste and visual cues seemed important. The red-tail's unenthusiastic response to the first mouse was based on its taste; her hesitation over the second was possibly a rapid association of white color and bitter taste.

On their next trial, the hawks were offered a dead unflavored black mouse (same color, different taste), again followed by an unflavored white mouse. The red-tail pounced on the black mouse immediately and consumed it eagerly after tasting its fur several times—with no flips of the head or flinging of the mouse, as had occurred with the bitter black and bitter white mice. Her actual consumption of this mouse was identical in appearance to her consumption of the unflavored white mouse that followed.

Once started, the rough-legged hawk also ate the unflavored black mouse in much the same manner as he ate the unflavored white mice. The bird seized the mouse immediately but carried it around the cage in his talons for a few minutes before tasting it. He then proceeded to tear off a strip of skin and sample the inside of the mouse. Apparently satisfied with what he tasted, he then bolted the entire mouse in one gulp.

The conclusion drawn from these series of tests was that while both taste and visual information are significant to raptors in rejecting a prey previously paired with illness, the taste of the prey animal is primary in guiding their post-treatment response to that animal as food.

Attention was next turned to examining the strength of the conditioned aversion as a function of time. The red-tailed hawk was given ten extinction trials on dead bitter black mice, one trial every four days. It is especially interesting that during these trials,

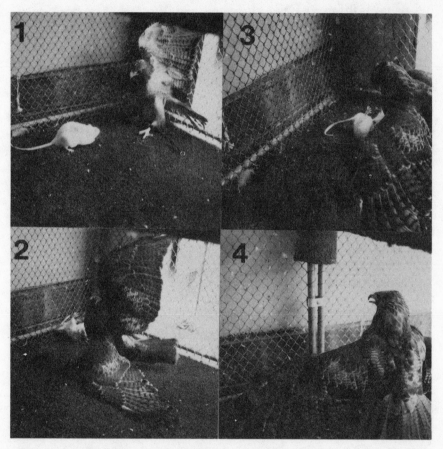

Figure 2.9 This sequence demonstrates the killing behavior of the female red-tailed hawk. The hawk confronts her quarry (1), then makes her strike (2). The hawk impales the rat with her taloned feet, immobilizing it with her grip (3). She then spreads her wings and tail feathers ("mantles-over") to shield her kill from intruders (4).

the bird's initial discriminative response to color of the mice offered was apparent. Latencies to pounce on the prey and contact it with her beak were approximately the same for both black and white mice throughout extinction trials, and these latencies were similar to those recorded before treatment. The bird's response to the taste cue, on the other hand, was remarkably long-lasting. The black mice were rejected by the hawk after she repeatedly tasted their bitter-soaked fur during the first three trials. On her fourth trial, the bird opened the mouse at the neck and ate a few small bites from the head. She tore off several pieces with her beak, then flung them away as if their taste was repellent, and picked up and tossed the body several times. Her behavior was similar during the fifth trial, except that she consumed the entire head. The red-tail proceeded to devour the bitter black mouse, except some small pieces of viscera, on the sixth extinction trial. Her consumption time was still far above baseline at this time, but extinction was complete by the tenth trial.

The rough-legged hawk has now been given more than thirty extinction trials with dead bitter black mice, but he ate the mice only on the twenty-third and twenty-fourth trials. This hawk demonstrated a stronger response to the color cue during extinction trials than did the red-tailed hawk. On most occasions, he does not taste the mouse and avoids it purely on the basis of color; at other times, he rejects the mouse after tasting it. This bird is obviously extinguishing far more slowly than did the red-tailed hawk. A possible cause of this difference may be that wilder hawks tend to be more suspicious of food offered them than most domesticated hawks.

We are currently examining conditioned aversions in hawks for whom the "unsafe" mice differ from "safe" mice along a single dimension, either color or taste. Although we are testing only one bird with each cue at the present time, we are obtaining results consistent with those just described in indicating that taste is more salient than color in food-illness conditioning. For instance, the red-tail in the "color-only" condition required three black mouse-illness pairings before she completely suppressed consumption of black mice. Moreover, she then generalized the aversion to white mice. This result suggests that we may have conditioned an aversion to taste of mice, though tests in which "safe" and "unsafe" mice are simultaneously presented indicated that she certainly learned something about color as well. She consistently chooses white mice over black ones but is very slow and "picky" when eating, regardless of mouse color. Our "taste-only" red-tail, on the other hand, completely suppressed consumption of bitter white mice after a single treatment, while continuing to eat plain white mice. Her disgust response to the unsafe prey is just as profound as that of the hawks tested with bitter black mice. This result is particularly encouraging, since before treatment this hawk showed no preference for unflavored mice over bitter mice in terms of time taken to eat them.

In conclusion, our experiments have indicated that establishment of taste aversions to specific prey readily inhibits consumption of that prey in coyotes, wolves, hawks, and rats. It also inhibits attack on the prey, though less consistently in the latter two species. Our preliminary field data suggest that an effective field program is feasible, and that future investigation could provide methods of establishing resident populations of pest species that will not consume valuable food commodities; or provide methods with which zoo personnel can manage the diets of captive animals exposed to the possible dangers of public feeding. Also, through the use of flavor-enhancement conditioning in which consumption of a flavor is followed by beneficial effects (Garcia et al. 1967), food-preference changes could be made for zoo animals whose diet must be altered due to lack of available special foods. Wild animals whose naturally preferred diets are endangered by ecological change could be conditioned to accept more prevalent foods and thus insure their own survival.

3 Social Dynamics and Communication

John C. Fentress, Rebecca Field, and Heather Parr *

Social behavior can be defined in terms of relationships between individuals within a given group. A basic assumption is that individuals within the group influence the behavior of other individuals within the same group. This is the essence of communication. We can say that communication occurs when the behavior of one or more individuals within a group is determined in part by the behavior of one or more other individuals within a group (cf. Altmann 1967; Hinde 1972; Wilson 1975). Thus the issue of communication is central to all considerations of social behavior.

With this as our starting point, we can question the means by which behavior of certain individuals within a group influences the behavior of other individuals. This approach leads to four separable questions: first, what are the factors that generate a given

* John C. Fentress and Heather Parr are associated with the Department of Psychology in the Life Science Centre of Dalhousie University, Halifax, Nova Scotia. Rebecca Field is associated with the Department of Mental Hygiene at The Johns Hopkins University, Baltimore, Maryland, and with the National Zoological Park, The Smithsonian Institution, Washington, D.C. Studies reported here were supported in part by the University of Oregon Graduate School, the National Institute of Mental Health, the National Research Council of Canada, and Dalhousie University. To Lupey, our instructor and friend, we dedicate this chapter.

class of communicatory behavior? Secondly, what are the details of
the signal used in communicatory behavior? Thirdly, what are the
factors that determine the response in animals that receive a given
signal? Finally, what are the means by which the "signaller"
modulates its behavior as a function of feedback from the "recipient"?

These four basic questions can be separated for the purpose of
analysis (Figure 3.1). It is particularly important, however, to
recognize that this dissection of the interconnected nexus of social
interaction is only for analytical convenience; eventually, we must
resynthesize our observations for an adequate picture.

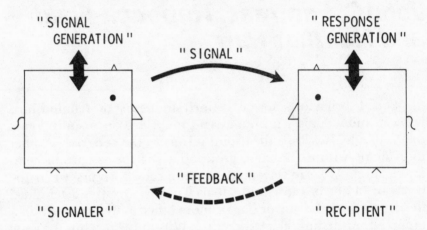

Figure 3.1 Schematic representation of four major questions in the study of social
communication: (1) factors intrinsic and extrinsic to the "signaler" that lead to
signal generation; (2) properties of the signal itself; (3) factors intrinsic and
extrinsic to the "recipient" which affect response to the signal; (4) "feedback"
influence of the recipient upon the signaler. While these dimensions can be
abstracted for analysis, they must be viewed as integral parts of the unified
communication system.

Captive studies have the great advantage of permitting close,
repeatable observations of social systems and communication
within a simplified framework. The value of these observations,
however, is limited to their demonstrated relevance to more natural, less controlled situations.

FRAMEWORK FOR ANALYSIS

To pursue studies of social behavior and communication in a

systematic manner, it is first necessary to outline some of the basic conceptual issues relevant to the study of behavior in a broader sense. One obvious question concerns the degree of interconnection between the behavior of a given individual within the group and specified factors that originate externally to this individual. To illustrate, communication cannot exist if the behavior of one individual is uninfluenced by factors which originate external to this individual; while, on the other hand, sophisticated communication systems depend upon *interpretation* of the exogenous signals as a function of the recipient's internal state. It is the *balance* between factors of intrinsic and extrinsic origin that our analyses must emphasize, both in terms of the generation of signals and the response to these signals.

A second class of questions concerns the balance between stabilities in social systems and the moment-to-moment dynamic fluctuations in individual behavior. This question of the relationship between stability and change in behavior lies at the heart of many investigations ranging from basic neurobiological processes to complex social structures (e.g., Fentress 1976a; Hinde and Stevenson-Hinde 1976).

A third class of questions concerns the balance between continuities and discontinuities in behavior. For example: if, at a descriptive level, we divide a wolf howl into its component parts, we must ask to what extent these parts are discontinuous from one another in terms of their vocal qualities. When we ask this same question in terms of causal processes which underlie descriptively distinguished components of behavior, the issues become more complex (e.g., Fentress 1973, 1976b).

Finally, it is essential to recognize that our questions can be framed at many different levels of analysis; and that, as a consequence of these different levels, interpretations concerning intrinsic or extrinsic factors, stability and change, and continuity versus discontinuity can vary significantly (e.g., Fentress 1976a, 1976b). These basic questions are summarized in diagrammatic form in Figure 3.2.

SOCIAL BEHAVIOR IN THE WOLF AS A TEST CASE

In this chapter, we shall concentrate attention upon problems of social dynamics and communication in captive timber wolves

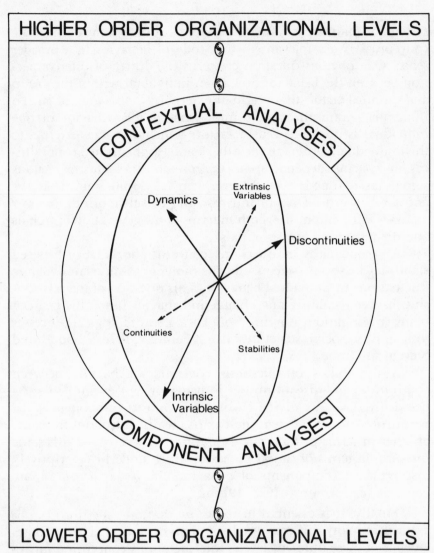

Figure 3.2 Major dimensions relevant to the examination of behavior at various analysis levels. Note that as one moves this "conceptual sphere" within the frame toward higher levels of analysis, factors initially viewed as contextual become basic components; extrinsic factors become intrinsic to the higher order system; and because of changed perceptual perspective, dynamic factors may appear more stable and discontinuities may appear more continuous. (Further explanation in text.)

(*Canis lupus*). We believe that the principles of social organization obtained through our work with these animals may be of interest, both in terms of the particular species and in terms of the more general issues of social organization and communication outlined. We should state at the outset that our work has travelled three complementary paths.

First, we have gained valuable insights by working closely with a small number of individual animals. These insights are often difficult to translate into formal objective terms; yet we feel they are important. Secondly, we have gathered formal data on aspects of social communication, particularly with reference to vocalizations, which provide a degree of detailed documentation not hitherto available. Thirdly, we have used both our informal interactions with these animals and our more formally acquired data to suggest relatively abstract strategies and principles that we feel can be applied to a variety of other species and situations.

A major goal of our investigations has been not just to indicate the most obvious and superficial features of social communication, but to gain a better appreciation for its subtleties. To do this, we have concentrated upon the related dimensions of *relationship* and *change* in social behavior. The idea of relationship between animals is obviously a basic defining characteristic of social behavior; but there are other aspects of the problem which demand attention such as the relationship between one form of behavioral expression and another as they occur over time and the relationship between a given signal and the broader context in which it occurs. The idea of change emphasizes the definition of behavior as a temporal process rather than as a static entity. "Snapshot" approaches to behavior inadequately represent the temporal flow of activities with which we ultimately must deal. Much of the literature on social communication and, indeed, on behavior in general, treats behavior as isolated packages frozen in time. Schenkel's (1947) classic catalogue of tail postures and facial expression in the wolf, for example, isolates individual expressions from one another and from their broader temporal context. As the performance of a musical composition can only be fully appreciated through the relationship of notes and pauses with respect to one another over time, so too might we expect "melodies" of behavioral expression to be understood most fully, both in terms of causation and of function, if we develop techniques for comprehending changing relationships—and relationships among changes—as they are expressed through time (see Fentress 1976b for a more detailed discussion).

A major contribution in this direction has been made by Golani (1976), who applied the Eshkol-Wachmann movement-notation system originated in conjunction with human ballet to the flow of social behavior in golden jackals and Tasmanian devils. Golani has demonstrated that careful description of movement flow from several explicitly defined criteria can reveal simplifying consistencies in behavior that are otherwise obscured. For example, two individual animals may go through quite complex series of movements which only become comprehensible when the observer determines rules for maintenance of actual or distant "contact" between the animals. In Golani's terms, social animals often interact as if they are connected together by an anatomical "joint." Through such descriptions, one can then pose more fundamental questions about mechanism and function; for example, what perceptual cues might animals employ in maintaining their distal contact?

One great advantage of captive studies is that they permit close observation not only of short-term interactions between animals such as those studied so beautifully by Golani, but also of formation and maintenance of long-term relationships between individuals within the social group (e.g., Hinde and Stevenson-Hinde 1976; Simpson 1976). As Hinde and Stevenson-Hinde point out, most investigations of social behavior have concentrated upon short-term interactions; while the role of long-term relationships in guiding these short-term interactions (as well as the other way around) is an issue of critical importance. This problem illustrates the difficulty of our task, for eventually we wish to say something about the relationship among relationships themselves (cf. Hinde 1972) over varying time scales for individuals and combinations of individuals within the social group.

In our own work, we have attempted to approach these problems by employing two complementary perspectives which we call *component analysis* and *contextual analysis* (see Figure 3.2). In component analysis, we attempt to subdivide a given behavioral phenomenon such as a vocalization series into meaningful component parts. Contextual analysis implies the opposite emphasis, for here we not only explicitly seek the relationship of these parts to one another over time, but also seek the broader context within which this behavioral series occurs (e.g., after a fight, before feeding, time of year or day). As the perspective of our analysis shifts, components can become context and vice versa (Figure 3.2); but the critical operational strategy of looking both "downstream" and "upstream" remains the same.

If these statements appear a bit formal and imposing in their demands, it is because we, as individuals who feel privileged for the opportunity of working closely with wolves for several years, are struck by the beauty and subtlety of their behavior. It is distressingly easy either to trivialize these animals as if they were mechanical robots performing to the tune of our own misconceptions, or to revel in complexity of the study—an act which borders upon praise of obscurity for its own sake. We are convinced that a middle ground exists, in which complexities of social behavior can be examined with precision.

When one surveys the existing literature, one finds that a major source of confusion in wolf social behavior stems from inadequate separation of descriptive, causal, and functional statements. For example, supposedly descriptive terms such as "imposing face," "self-confidence," and "submission" (Schenkel 1947), or "friendly," "submissive," and "aggressive" (Zimen 1972) have been primarily employed as descriptors; but these terms carry obvious, untested implications about both causation and function. Schenkel confronts the important question of submissive behavior *function* in wolves by referring to the *observed* distinction between "conflict" and "fighting." Since, however, descriptive documentation of "conflict" and "fighting" is not provided, apparently some interpretations are used to explain others without any clear referent to actual behavior. Clearly, even the adequate description of behavior that we interpret as social demands more careful attention.[1]

---

1. We are grateful to our colleague G. Moran for discussion of these issues.

COMPONENTS AND CONTEXT
OF WOLF VOCAL RESPONSES

If observations of visitors to a zoological park provide any indication, it is apparent that wolves and howling are commonly associated in the public mind. But, detailed investigations of vocalizations in wolves are relatively rare. Murie (1944) noted seventeen instances of howling in the animals he studied at Mount McKinley Park in Alaska. In eight of these instances, the animals had either been separated from the pack or had recently returned to a vacated home site. Mech (1966), Rutter and Pimlott (1968), and Peterson (1974) provide evidence that howls serve the function of assembling members of the same pack. Theberge and Falls (1967) studied captive wolves and found that howling increased when individuals were separated from their peers. The same researchers also provided evidence that wolves can discriminate between tape-recorded howls and live howls, and between live "howls" emitted by different human investigators (as measured by responsiveness to these "howls").

One suggestion is that wolves howl partly as a function of territorial maintenance. For example, Murie (1944), Mech (1966), Joslin (1966), and Theberge (1966) note that the approach of human investigators can generate howls accompanied by barks. Not all howls are thus accompanied, however, which indicates that, as a category, howling may be relatively heterogeneous. For example, Crisler (1958) and Zimen (1972) report chorus howls without accompanying barks in captive animals that are not obviously disturbed by extrinsic stimulation. Wolves may also respond to howls generated by individuals outside the pack, which supports the contention that howling may serve an important territorial function in addition to promoting group cohesion (e.g., Pimlott 1960; Joslin 1967; Harrington 1975).

Harrington (1975) has obtained recordings of the detailed structure of howling in the field. He has also obtained evidence for the differential receptivity of wolves to recorded wolf howls and human howls as a function of the broader context in which these howls are presented. He provides, for example, evidence to suggest that presence of pups with adults, presence of a recent kill, breeding season (February-March), social rank within the pack, and size of pack all influence the responsiveness of wolves to standardized howling presentations.

Harrington's data clearly suggest that the broader context within which a given howling stimulus is received is an important determinant of an individual's responsiveness to this howl. Zimen (1972) reports an increase in howling responsiveness of captive wolves during the breeding season; and Klinghammer (1975) observed that these seasonal fluctuations in responsiveness to howls depend in part upon the animal's social rank within the pack; i.e., in Klinghammer's captive group of animals, most increased howling during breeding season could be attributed to the alpha male. This situation appears generally applicable to other wild canids. For example, Skead (1974) found that black-backed jackal males call significantly more often during the three months when females are in estrus; and Young and Jackson (1951) report a greater tendency for coyotes to howl during the breeding season. We too have obtained evidence that both spontaneous howling and howling in response to presented howl recordings in captive timber wolves peak during the breeding season.[2]

While these studies conclusively demonstrate that responsiveness to a given social signal partially depends upon the context of that signal's reception, they do not directly answer questions about (1) responsiveness to signals of different quality, (2) differences in responsiveness between individual animals apart from obvious differences in social hierarchy, or (3) short-term changes in responsiveness that can be reversed in a matter of minutes, hours, or days. With specific regard to the question of context as a determinant of responsiveness to a given signal, Smith (1968) has provided a useful distinction between *message* and *meaning* of a given signal on both long-term and short-term bases. Yet the evidence on short-term changes in responsiveness to a standardized signal of obvious social significance is minimal to date.

Shalter, Fentress, and Young (1977) have studied the effect of context upon responses of two female wolf pups to standardized recordings of vocal signals. The animals were housed in a 1 by 1.2 by 1.5-meter pen in the same room where the stimuli were presented. Six acoustic stimuli were employed (Figure 3.3): (1) noise,

---

2. This review of data borrows heavily from Harrington (1975).

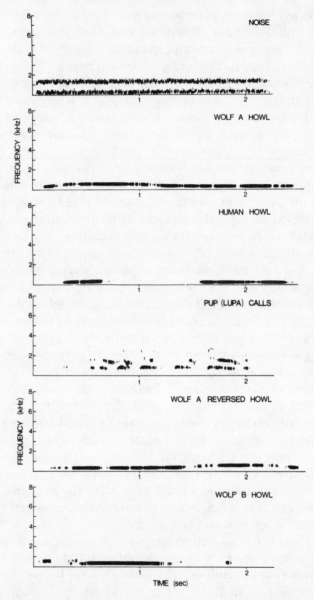

Figure 3.3 Spectrograms of the six acoustic stimuli used to determine factors which contribute to vocal responses in young wolves (from Shalter, Fentress, and Young 1976).

(2) adult wolf A howl, (3) human howl, (4) a pup's vocal responses to wolf A howl, (5) wolf A reversed howl, and (6) adult wolf B howl. Mean duration of these signals was just over 2 seconds (range 1.8 to 2.4 seconds), with a standardized amplitude of 85 ± 2 decibels; re 20μNper square meter as measured at the center of the pups' pen (B weighting curve, General Radio 1565-B sound-level meter). Each vocal stimulus was presented repeatedly for 30 seconds (presentation 1), followed 2 minutes later by another 30-second presentation (presentation 2) of the same stimulus. The reception context of each signal was modulated by introducing a human observer, a dog, or live mice on specified days prior to, and/or during, stimulus playbacks. Video and audio tapes were used to document in detail the responses of the wolf pups to the experimental stimuli, and also to gather more general data on the ongoing behavior of these animals. In this way, it was possible to document five major factors that can affect an animal's response to social signals: (1) the nature of the stimulus, (2) the number of repetitions of a given signal, (3) the influence of contextual variables on responsiveness, (4) ongoing behavior of the animals, and (5) individual differences in responsiveness of the animals. For the purpose of simplification, we shall concentrate upon vocal responses of the animals.

One of the pups (Kusa) responded vocally during only 13 of the 200 stimulus presentations, while the other animal (Lupa) responded vocally to 143 presentations of the auditory stimuli, thus indicating fundamental individual differences. Moreover, Kusa *never* howled in response to the experimental stimuli, whereas Lupa not only howled in response but also emitted the typical squeaking or whining sounds (the only sounds recorded from Kusa). Factors influencing total vocal responsiveness of each animal appeared similar, however. Figure 3.4 summarizes total vocal response of the two animals to different auditory stimuli as a function of time and contextual manipulations.

It is clear that neither animal responded vocally to the noise stimulus, even though this stimulus was initially novel and had an overall similarity to the other stimuli in terms of amplitude, duration, and frequency range. Further, neither pup responded as vigorously to recordings of Lupa's vocalizations as they did to

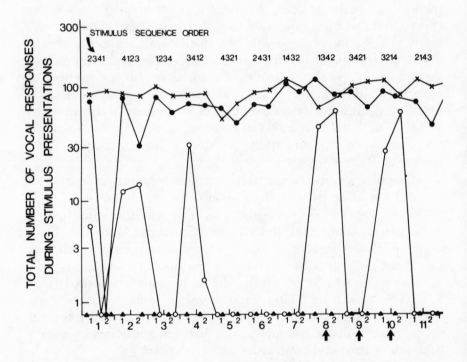

Figure 3.4 Graphs indicate the number of vocal responses during presentations 1 and 2 of the acoustic signals for each of the 24 days of the experiment. Stimulus sequence order is indicated for each day; the numbers are identified on right.

78

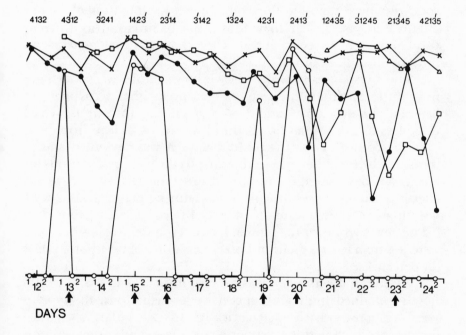

Arrows beneath the figure indicate days on which the context for stimuli received was manipulated (here combined for human observer, dog, and caged mice). (From Shalter, Fentress, and Young, 1976.)

howls by wolf A or B. Only in three of the sixteen presentations in
which there was vocal response to Lupa's vocalizations did this
response exceed that triggered by the human howls; and in only
two (12.5 percent) cases did the vocal response to Lupa's vocalizations surpass the response elicited by reversed howls. Figure 3.4
illustrates that the recorded howls of wolf A elicited a greater
number of vocalizations than did the recorded human howls in 45
of 48 comparisons (94 percent).

On its initial presentations (days thirteen through seventeen),
the reversed howls elicited an even greater number of responses
than did the same howl (howl A) played in a forward direction.
Subsequently, however, response to the reversed howl dropped well
below response to the forward howl. Similarly, during its initial
presentations, the wolf B howl also evoked a greater number of
vocal responses than did the wolf A howl. Examination of data in
(Figure 3.4 makes clear that differential habituation to different
vocal signals may occur, and that novel signals may generate an
initially high level of response.

Vocal responses to wolf A howl, for example, were strikingly
resistant to habituation effects, as measured either across days or
between presentations 1 and 2 on a given day. In contrast, vocal
responses to the human-howl stimulus showed a clear drop, both
over the course of the twenty-four days it was presented and as
measured by the difference in response to the two presentations for
any given day. To illustrate the distinction between short-term
effects, vocal responses to wolf A howl during presentation 2 were
less than for presentation 1 on only fifteen of twenty-four days
(62 percent), while for the human howl, responses to presentation 2
were less than to presentation 1 on twenty-three of twenty-four days
(96 percent).

Although the *number* of vocal responses to wolf A and wolf B
howls remained high with successive presentations, the *type* of
vocal response varied systematically. In particular, a marked
decrement in howling responses to these and other stimuli
occurred over the twenty-four days of the experiment; and for the
wolf A howl, as well as for the reversed howl and human howl,
there was marked reduction in howling responses to second
presentation of the stimulus on any given day. This drop was

respectively seen in fifteen of eighteen, six of seven, and five of five cases. It is apparent that captive studies can thus be used to analyze processes of stimulus specificity and response specificity, here illustrated within the context of habituation, with a degree of precision that would be difficult, if not impossible, in the field.

Captive studies also permit systematic manipulation of the environmental context within which a signal is received. For example, of the thirteen stimulus presentations during which Kusa responded vocally, ten (77 percent) occurred with the human observer present or immediately after a preceding contact with the dog. As illustrated in Figure 3.4, vocal responses of Lupa to her own recording were greatly enhanced by manipulations of the environmental context (indicated by arrows in the figure). If one assumes that presence or absence of vocal response on a given day was independent of response patterns found on the previous day, the augmenting effect of contextual manipulations upon vocal responses to the Lupa stimulus is highly significant (i.e., the effect occurred in the correct direction in twenty-nine of thirty-five cases after habituation on days eight through twenty-four; $p < .001$, binomial test). Obviously more subtle influences of context exist which may contribute to variations in response to a given signal, both in captivity and in the field; and it will surprise no one who has worked closely with wild animals that they adjust their social behavior as a combined function of a variety of factors. What these experiments indicate is that the underlying network of causal relations can be subjected to direct empirical testing under appropriately controlled conditions.

CLASSIFICATION OF VOCAL SIGNALS

As indicated in Figure 3.1, the precise description and classification of communication signals and the context within which they are elicited is a critical task. Here we shall concentrate upon the structure of vocal signals, analysis of which has been greatly aided by technological advances such as the sound spectrogram (e.g., Thorpe 1963) and methods for digital analysis (Field 1976). Like sheet music, spectrograms represent changes in frequency with changes in time (cf Figure 3.3). While possible distortions of sound in spectrographic representation must be watched for (cf.

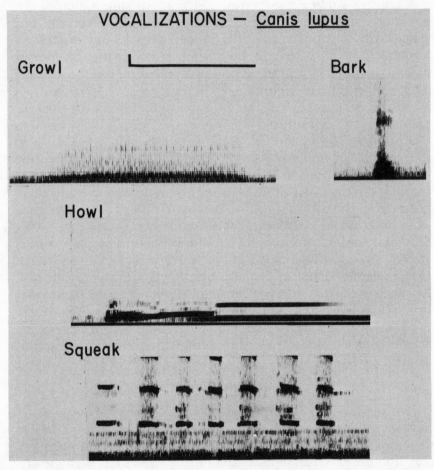

Figure 3.5 Spectrograms illustrating four basic categories in the vocal repertoire of timber wolves. As in subsequent spectrograms, wide-band filters with a frequency range of 80– 8000 hertz were used. Vertical scale represents 1 kilohertz and horizontal scale represents 1 second. (After Field 1975.)

Greenwalt 1968; Marler 1973; Eisenberg 1976), spectrogram analysis remains a prominent method in studies of vocal behavior and has been used in the present analysis of the vocal repertoire of wolves (cf. Fentress 1967; Theberge and Falls 1967; Harrington 1975).

Wolf vocalizations have usually been divided into four basic groups (Figure 3.5), (1) growls, (2) barks, (3) howls, and (4) squeaks or whines (cf. Joslin 1967; Fentress 1967)—although Tembrock (1970) divided the repertoire into eight discrete categories. Yet, in most studies of wolf vocalizations, minimal use of sound structure measurements in defining sound classes makes it difficult to determine whether sounds reported by one investigator are included in a category established by another.

The spectrograms in Figure 3.5 are samples of sounds within each category. Although discontinuities within vocalizations often occur which allow us to distinguish between different classes of vocalizations, there is sufficient variability within each group to dictate more critical evaluation of both sounds and contexts within which the sounds occur. Questions of variability and discreteness of vocal signals have been of major concern in studies with other groups such as the primates (e.g., Marler 1965; Green 1975; Gautier 1974), but firm conclusions are not yet possible for the vocal repertoire of wolves. As indicated in Figure 3.2 answers to these questions are partially dependent upon the total conceptual framework, as well as upon levels of analysis employed in any given study; and any unitary judgments must be regarded with caution at the present time.

Measures of spectral characteristics and variability for one category, that of squeaks, have been reported elsewhere in more detail (Field 1975). This group of vocalizations are particularly well-suited for study in captivity conditions, since qualities of high frequency and low volume mean that squeaks do not travel far and thus are easily missed in field studies. Also, because squeaks are close-range sounds, they are probably involved in controlling and coordinating interactions within the social group. Marler (1972) has found that close-range sounds in primate groups are highly graded in structure, but accompanying expressive movements may serve to clarify the variable sounds.

If we broadly define squeaks as high-frequency, tonal sounds, we may also find in their structure a considerable amount of variability which seems to parallel changes in the accompanying context. For instance, squeak sounds emitted during a howling session (Figure 3.6) have long syllable duration and considerable frequency fluctuation. In a preliminary examination of associations between sound structure and context (Fentress and Field-Lockhart 1974), we found that squeaks from one individual varied in form as the context changed (Figure 3.7); yet squeaks emitted by different wolves on separate occasions within similar contexts appeared strikingly similar in form (Figure 3.8). Even when this comparison was made between adult females of separate but closely related species (Figure 3.9), similarities in squeak form were apparent. The important point to understand at this juncture is that different classes of sounds are generally distinguishable from one another (i.e., discontinuous), and that variability within a class may be context-dependent.

While these results suggest a positive correlation between sound structure and context, they represent small sample sizes from which "representative" sounds were selected. Also, while similar sounds appeared in similar contexts, differences in structure were also seen. What were the sources of these dissimilarities (e.g., individual, age, sex, or species differences)? To approach this question, it is necessary to measure sound structure and variability within those measures in such a way that comparisons are feasible within and between these different clusters of animals.

Therefore, in the second stage of this research (Field 1975), measurements were made of syllables within squeak phrases (i.e., component analysis) emitted by the adult male, younger adult female, and two wolf pups approximately one-and-a-half months old (Figure 3.10). A syllable, as defined here, is a continuous trace on the spectrogram; while a phrase is a group of syllables separated from other groups by an interval longer than any interval between syllables. The context in each case was the approach of a human "friend" to the vocalizing wolf. Whole phrases were used in the analysis because we found that syllable characteristics may change throughout a phrase (see Figure 3.11), making a sampled group of syllables not necessarily representative of a whole phrase.

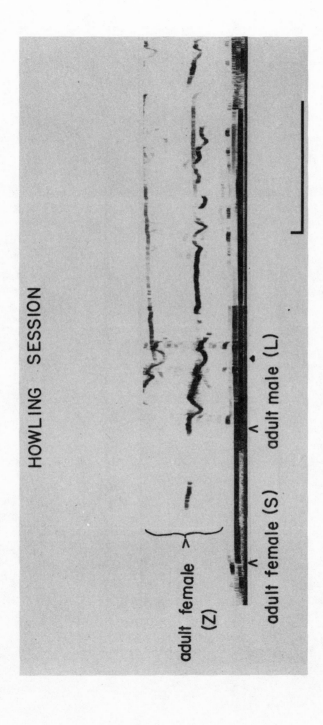

Figure 3.6 Squeaks by adult female (Z) during howling session (adult female, S, and adult male, L). (After Field 1975.)

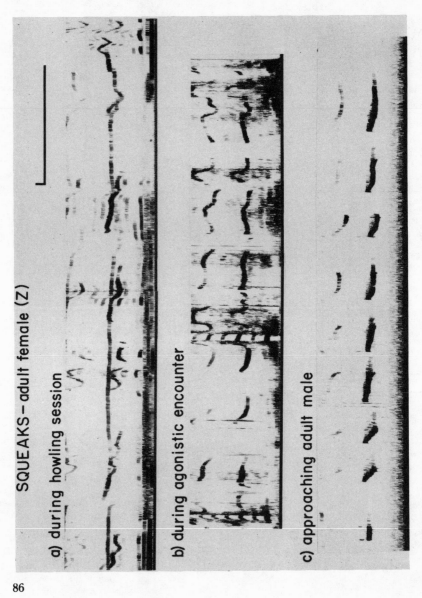

Figure 3.7 Partial squeak phrases by same female wolf in three different contexts. (After Field 1975.)

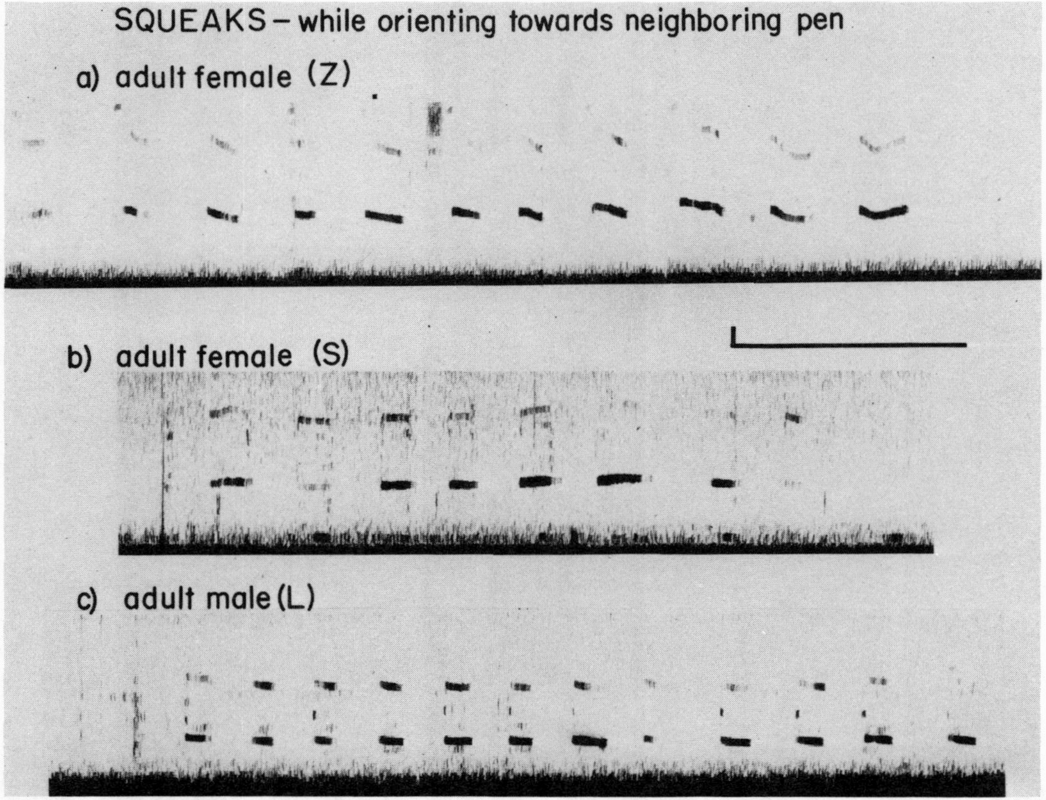

SQUEAKS – while orienting towards neighboring pen

a) adult female (Z)

b) adult female (S)

c) adult male (L)

Figure 3.8 Partial squeak phrases by three adult wolves in similar context. (After Field 1975.)

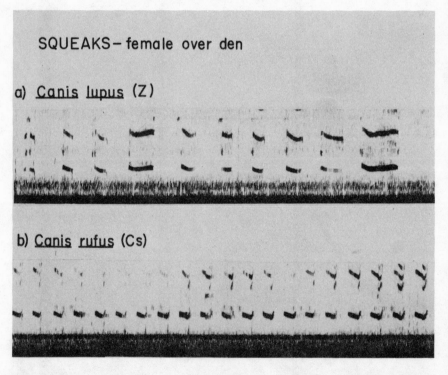

Figure 3.9 Comparison between squeak phrases of adult female timber wolf and adult female red wolf-coyote hybrid (suspected) while standing over a den containing pups on separate occasions. (After Field 1975.)

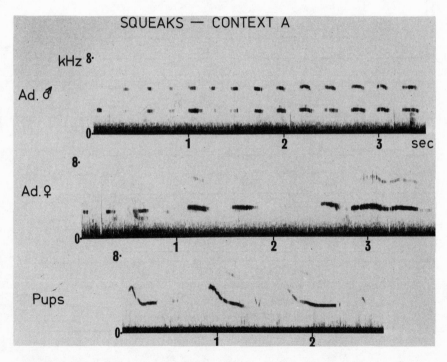

Figure 3.10 Spectrograms of partial squeak phrases recorded from adult male and adult female timber wolves and from one of two wolf pups during similar contexts of distance reduction between vocalizer and a human handler. (After Field 1975.)

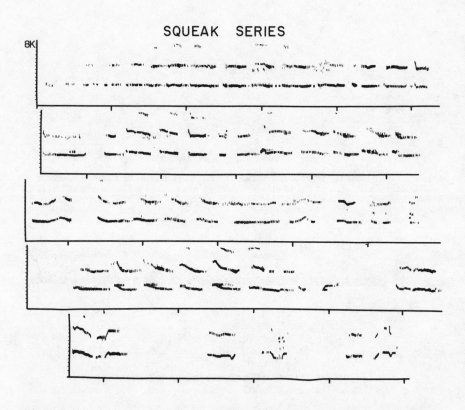

SQUEAK SERIES

Figure 3.11 An extended phrase of squeak syllables given while an adult female timber wolf, carrying chicken, was chased by two yearling wolves. In this redrawn spectrogram, each line is a temporal continuation of the one before it.

90

Using a digitizer for obtaining measurements from spectrograms (Field 1976), we obtained mean values from the phrases. Mean values included those for syllable duration, syllable frequency, interval duration, and frequency fluctuations (or range). Values for pairs of these measures for each phrase could then be plotted. As a representation of variability among phrases, ellipses were constructed around a grand mean for each group of phrases and one standard deviation on either side of the grand mean for each measure. With these "variability ellipses," groups of phrases are comparable not only in relative mean values but also in relative variability for the measures involved.

Using these methods of illustration, we compared squeaks vocalized on separate but similar occasions by the adult male, adult female, and pups. Because details of this research have been reported elsewhere (Field 1975), only highlights of results will be mentioned here.

Squeaks emitted by the adult female wolf were nearly twice as long in mean syllable duration and were higher in mean syllable frequency when compared to squeaks from the adult make (Figure 3.12). Variability in mean syllable duration was similar for the two adults' squeaks (male ± 0.040 seconds; female ± 0.054 seconds), but the female's squeaks were more than three times more variable in mean syllable frequency than the adult male's.

Comparisons between adult squeaks and pup squeaks (Figure 3.13) for the context in question showed that pups had higher mean syllable frequency and longer mean syllable duration (although squeaks from the adult female and pups were close on the latter measure: female 0.211 seconds; pups 0.228 seconds). Pups' squeaks were almost twice as variable in mean syllable duration, but were similar in variability of mean syllable frequency (pups ±0.261 kilohertz; adults ±0.215 kilohertz).

Similar relationships existed between the three groups of squeak phrases in other measures, such as mean syllable frequency fluctuation versus mean syllable frequency (Figure 3.14). But other combined measures, such as mean interval duration versus mean syllable duration, produced different relationships. In Figure 3.15, the adult female's squeaks were longest in mean interval duration and greatest in variability for that measure. The male's squeaks were shortest in both mean interval duration and mean syllable duration.

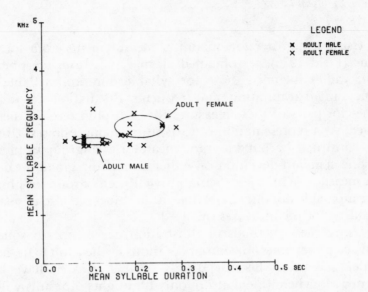

Figure 3.12 Comparison of squeak phrases given by adult male and female timber wolves and by male and female timber wolf pups on separate but similar occasions. Squeaks from adult male and female wolves are compared on measures of syllable duration and frequency. (After Field 1975.)

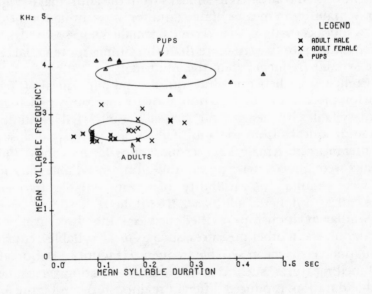

Figure 3.13 Adults' squeaks are compared with pups' squeaks on the same measures. (After Field 1975.)

92

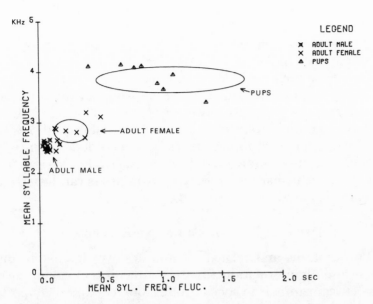

Figure 3.14 Squeaks from both adults and the pups are compared on measures of mean syllable frequency and frequency fluctuation. (After Field 1975.)

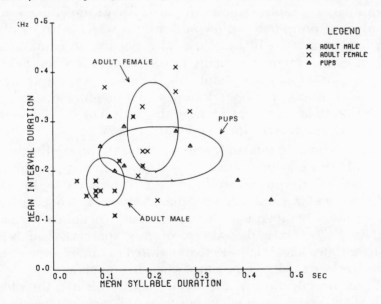

Figure 3.15 The squeaks are compared on measures of mean interval duration and mean syllable duration. See text for description of ellipse construction. (After Field 1975.)

The squeaks analyzed here were only those given by a few animals in one type of context. Also, only a few measures of sound structure have been used. These initial data, however, indicate the possibility of applying quantitative measures to the basic taxonomic issue of distinctions between vocalization "types" as well as between variations within these "types." "Variability ellipses" can be compared for vocalization "types," contexts of occurrence, age, sex, social position, individual differences, etc. The great value of captive studies is that the underlying parameters can be separated for detailed analysis.

DYNAMIC RELATIONSHIPS AND EXPERIENCE

Vocalizations and related communicatory signals obviously provide useful information for the moment-to-moment interactions between animals in social groups; but it is equally important to seek other measures for longer-term relationships and their fluctuations in time (cf. Hinde and Stevenson-Hinde 1976). Here the advantages of captive studies are particularly compelling, since not only is it possible to follow in detail the social behavior of a given animal over its lifetime; it is also possible to manipulate experiences that might contribute to the changes or stabilities of social relationships. For example, the behavior of a single male timber wolf (Lupey) has been documented from four weeks of age until his death at eleven years within the context of varying social opportunities and restrictions (cf. Fentress 1967).

The first goal of this study was to explore the possibility that a successful relationship between wolf and human could be established—if, in contrast to the approach employed by Kramer (1963), particular emphasis was placed upon a "cooperative" relationship rather than upon strict dominance of animal by man (Fentress 1967). Once the success of this approach had been established, two female cubs were introduced to Lupey—one when he was five years old and the other when he was six.

Upon introduction of the second female (Zelda), the older female (Sylva) established what most observers would agree was a strict dominance relationship over the younger animal. For example, Sylva would frequently step between Lupey and Zelda, and during subsequent breeding seasons she would repeatedly pin Zelda to the ground.

Through the spring of 1973, when Lupey was ten years of age, no successful breeding had occurred. Occasionally Lupey would display preliminary tendencies to mount one of the two females, but on most occasions any approaches he made toward Zelda were truncated by the active interference of Sylva.

We could establish a correlation between Lupey's intention-mounts and the immediately preceding agonistic encounters between the two females; i.e., shortly after a confrontation between the two females (in which Sylva was invariably victorious), Lupey would often approach one or the other female and perform preliminary mounting attempts. The possibility that fighting between female wolves increases likelihood of male mounting attempts deserves further examination.

For a variety of reasons, the older female (Sylva) was temporarily removed from the social group during the breeding season of 1973. Upon her reintroduction, the two females engaged in a prolonged fight of more than an hour in duration (Figure 3.16); and Sylva was finally removed for her own protection. Up until April 1977, when Sylva died, the two females continued to challenge one another across a fence barrier; their relationship had clearly changed during the course of their short separation.

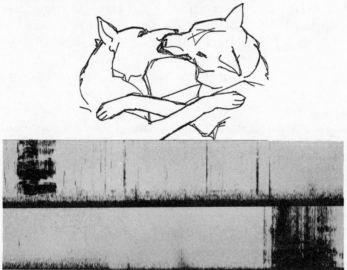

Figure 3.16 *Upper,* an upright jaw-locking posture used by two adult female wolves during agonistic encounter. *Lower,* explosive barklike sounds, separated and terminated by silence, which ended prolonged period of squeaking and were followed by overt fighting.

In the fall of 1973 several months following the fight between the females, two three-month-old female pups were introduced into the pen with Lupey and Zelda. Although neither of these adult animals had bred, they spent much time feeding the pups and digging dens. During the subsequent spring (1974), a litter of six cubs was born. A possible but presently speculative factor in the successful breeding was the prior care given the pups by the adults.

The clearest example of changes in social relationship over time has been provided by our colleague J. Ryon, whose data concerns the feeding patterns displayed by the two adult animals (Lupey and Zelda), the introduced cubs (Suzie and Sally), and the newborn litter over the subsequent breeding season. (Summary data are presented in Figure 3.17. Suzie and Sally are here designated as yearlings.)

Figure 3.17 displays representative data collected over 102 observation nights. The figure is broken into three horizontal sections which indicate feeding patterns of the adult male (L), the adult female (Z), two yearlings (SS), and the cubs or pups (P) over three different time periods. The first period of forty-six observation nights, from August 29, 1973 through December 1, 1973, came before the pups were born and before onset of breeding. The second period of twenty-one observation nights combines data just prior to the April 19, 1974, birth of the pups and approximately four weeks after their birth. (Data within this period were homogeneous for the comparison of days prior to, and following, birth of the pups.) The third period of thirty-five observation nights covers the four-month period after the pups were first seen out of the den ("above ground").

Data for regurgitation and carrying of food are represented by the numbers enclosed by squares and parentheses in the left half of Figure 3.17. Each of the three columns in the left side represents animals seen to regurgitate and/or carry food in their mouths to other members of the social group. The nine data blocks within the figure's left side break down the feeding patterns of animals which received food. Thus, for example, the upper left data block indicates that the male (L) was not observed to regurgitate food to the adult female (Z) between August 29, 1973 and December 1, 1973; but that he was observed to regurgitate food to the yearlings (SS) on thirty-six occasions. Also during this period, the adult male (L) was never observed to carry and deposit food to the adult female (Z); while, on seventy-five occasions, he was observed to carry and deposit food to the yearlings (SS).

	REGURGITATION ☐ CARRY ( )			SOLICIT FEEDING		
	MALE (L)	FEMALE (Z)	YEARLINGS (SS)	FEMALE (Z)	YEARLINGS (SS)	PUPS (P)
BEFORE LITTER (AUG. 29/73– DEC. 1/73) [N=46]	Z[0] (0) SS[36] (75)	SS[11] (29)		L[5]	L[61] Z[34]	
LITTER *APR. 19/74 (APR. 1/74– MAY 13/74) [N=21]	Z[16] (0) SS[8] (26)	SS[0] (0)		L[22]	L[16] Z[2]	
LITTER ABOVE GROUND (MAY 14/74– SEPT 14/74) [N=35]	Z[1] (0) SS[11] (10) P[25] (8)	SS[0] (5) P[15] (11)	P[19] (7)	L[7]	L[3] Z[0]	L[12] Z[10] SS[22]

N = OBSERVATION DAYS

Figure 3.17 Patterns of feeding and food-solicitation among a group of wolves before the breeding season (August 29– December 1), shortly before and after the birth of pups (April 1– May 13), and after the pups emerged from the den (May 14– September 14). Each column indicates the animal that fed the others (left half of figure) or solicited from the others (right half of figure). *Left*, note separate entries for feeding by regurgitation (squares) and carrying of food in the mouth (parentheses).

Several clear patterns of feeding behavior emerge from these data. Prior to the breeding season, for example, Lupey (L) regurgitated and carried food to the yearlings (SS), but was never observed to do so for the adult female (Z). The ratio of carrying to regurgitation for the yearlings was approximately 2:1. Shortly before and after birth of the adult female's pups, Lupey was observed to regurgitate to her on sixteen occasions. He was never observed to carry and deposit food before the adult female, although he continued this practice for the yearlings (i.e., a ratio of twenty-six observed carries and deposits to eight observed regurgitations). Lupey clearly had different styles of feeding for the pup's mother than for the yearlings. Also during this period, he was observed to regurgitate food twice as often to the adult female as to the yearlings, in striking contrast to his behavior prior to the breeding season. Once the litter was observed above ground (outside the den), Lupey's feeding patterns again shifted dramatically. During this period, he spent considerable time feeding the newborn pups and almost ceased his feeding of the adult female. The pups were observed to be fed three times as often by regurgitation as by Lupey's carrying and dropping of food. This ratio is strikingly different from those of the previous two time periods. Notice at this last stage that Lupey also appeared to feed the yearlings equally often by regurgitation and carrying, in contrast to the previous time periods. This observation suggests that not only had his feeding ratio of different animals changed, but also the style with which he fed these animals. This shift between forms of behavior over time, combined with relatively stable forms of expression within a more limited time frame, indicates to us the utility of thinking about social behavior within the general context of dynamic-stabilities (cf. Hinde and Stevenson-Hinde 1976).

This is but part of the picture defined at the level of total "social nexus." For example, the second column of the Figure 3.17 indicates that the adult female (Z) was observed to feed the year-lings (which were not her pups) frequently between August 29 and December 1,1973, but less frequently after that period. When the pups (P) emerged above ground, she regurgitated and carried food to them moderately often, but in total not as often as did the adult male (L). Most interestingly (column 3 of Figure 3.17), the year-lings frequently regurgitated food to the pups after they emerged from the den. Often the yearlings were observed to solicit food from the adult male, obtain it, and then feed the pups.[3]

---

3. Feeding of the pups is recorded only for cases where it could be directly observed. Thus, while the female (Z) was observed to carry food into the den before the pups emerged above ground, these data are omitted from the present table.

The second half of Figure 3.17 concerns the number of times animals were observed to solicit feeding from other animals. It is clear, for example, that the adult female (Z) increased food solicitation from the adult male (L) shortly before and after the pups were born. At the same time, the yearlings decreased their solicitation of feeding from the adult female. Once the pups had emerged from the den, they were observed to solicit food as much from the two yearlings combined as from the two adult animals combined. During this period the yearlings had reduced their solicitation of food from either adult. The definition of "signal" and the distinction between "signaller" and "recipient" become moot (above).

These data indicate both the close interconnection between dimensions such as feeding and food solicitation, and the dynamic stabilities of social organization which appear upon examination of different time frames (i.e., before and after the birth of new members of the social group). It is important to recognize reciprocal interactions between the behavior of different animals within the context of the group's current social structure. For example, there is a relatively good, but not perfect, correlation between feeding and soliciting behavior of different individuals. Thus, feeding of the adult female (Z) by the adult male (L) correlated relatively closely with patterns of food solicitation by the adult female, whereas the correlation between feeding by the adult male of the yearlings and their patterns of active solicitation seem less tightly connected. It is at this level of social organization that we need more precise formulations of behavioral control.

Many intricacies of social communication emerge from observations of the present type. For example, food solicitation by the adult female (Z) was actively rebuffed by the adult male (L) both before and after the interval when the pups were born. This rebuff was likely to affect the female's solicitations, which in turn affected the number of times she was fed by the adult male. The point here is that communication networks are obviously complex, and we may gain fuller insight by extending our framework of observation in addition to refining our analyses within a more narrowly defined framework.

From the perspective of analyzing the more subtle parameters of social behavior, we are not particularly predisposed to simple "lock and key" models of communication. Even in highly complex social animals such as wolves, however, traditional ethological

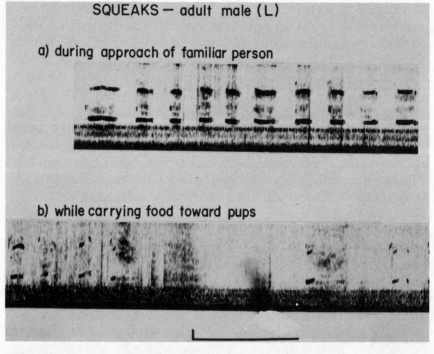

Figure 3.18 Comparison of squeaks given by male wolf during approach of familiar person and occasionally recorded brief-irregular squeaks given by the same animal when carrying food toward pups.

"lock and key" models appear to have some merit in suggesting questions for future inquiry. To illustrate, we were struck by the dual fact that Lupey's vocalization patterns prior to regurgitation were often different from similar squeaking sounds used by him in other contexts (Figure 3.18), and that regurgitation itself often depended upon preceding "face nuzzling" of Lupey by the pups. On several occasions, we pre-fed the pups and then gave food to Lupey. He would then take his rations to the pups, accompanying this action with characteristic squeaking sounds (Figure 3.18). Since the pups had already been fed by us, they failed to respond to Lupey's squeaks with the usual pattern of face nuzzling. Under these conditions Lupey frequently failed to either regurgitate or drop carried food, as if stimulation from the pups played an important role in the completion of this behavior sequence. Further studies are needed to untangle these factors in mammals.

While shifts in social relationships as measured by regurgitation patterns are striking, we must be careful not to underestimate the phenotypic stabilities of behavior that can be revealed in captive studies. The shifts in feeding relationships that we have found may reveal important characteristics of wolves in the field.

SOCIAL ARCHITECTURE

The construction and use of dens play critical roles in the care of pups in the field (e.g., Young and Goldman 1944; Mech 1970) and here J. Ryon (Ryon, 1977) has provided additional evidence that denning behavior can be successfully examined in captivity (Figure 3.19). Dens were dug by the wolves in April and May, 1972 and 1973, and collapsed during subsequent winter rains. No pups were born in 1972 and 1973, and our measurements of these dens were approximate.

Three more dens were constructed between March and October, 1974, the period which encompassed the birth of the six pups noted above. Digging of the first den started a month before parturition, and was completed in less than a week. The other two dens were begun at the same time, after the first was completed, and the excavation continued until shortly after the pups were born. In each case the adult male was observed to do most of the digging, the yearlings did very little, and the adult female an intermediate amount.

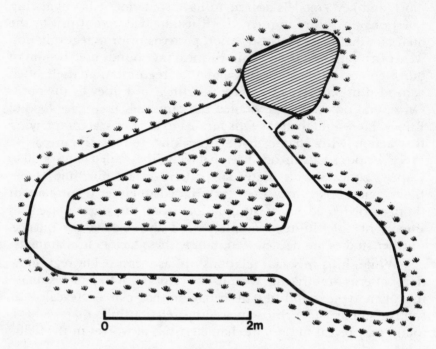

0   2m

Figure 3.19 Plans of timber wolf den from above, plus elevation and three transverse sections. Measurements by J. Ryon.

Initially, digging consisted of alternate scratching movements of the forepaws. As the tunnel deepened, simultaneous backward movements of both forelimbs became more common. Internal chambers were shaped by swiping movements of one forepaw, often with the animal lying on its back or side. When a pile of earth had accumulated, it was pushed backward with the side of the forelimbs, then removed from the tunnel by digging movements. The longest periods of digging were in the early morning and evening.

After the animals were transferred to Dalhousie University, Ryon excavated and measured the dens. For each of the three dens measured, the entrance tunnel (138–183 cm long, 36–41 cm wide, 31–38 cm high) sloped into the ground at an angle of about 45° (range 43°–48°) and to a depth of approximately 0.5 meter. A horizontal tunnel (10–312 cm long, 36–61 cm wide, 28–30 cm high) was constructed below the surface at that depth, and a terminal chamber (45-127 cm long, 70-117 cm wide, 36 cm high) with a domed roof and slightly dished floor was formed. The entrance of the tunnel terminated upon a transverse channel about 5 cm deep, bordered on the inner edge by a slightly raised dam. Ryon (1977) cites other details concerning the similarities and differences between these dens, and should be consulted for further information. While her data surpass in detail previous observations, the basic forms she describes resemble those of dens found in the field (e.g., Murie 1944; Cowan 1947; Criddle 1947; Jordan, Shelton, and Allen 1967; Haber 1968).

Den construction and use are particularly interesting in the study of social behavior since dens are an integral part in the rearing of pups. That den digging as well as other forms of "parental" behavior can be documented for animals reared under restricted environmental conditions suggests that these may represent strong phenotypic stabilities in social behavior which can be revealed by careful documentation of animals in captivity.

BOUNDARIES OF SOCIAL RELATIONSHIPS IN PERSPECTIVE

Thus we return to a more general consideration of the boundary conditions (Fentress 1973, 1976b) that underlie social behavior in complex animal species. There is no doubt, for example, that changes in environment can produce profound alterations in the details of social relationships. We have recently had opportunity to examine changes in the behavior of hand-reared wolves toward human observers as a function of placing these animals into a larger enclosure (i.e., into 10 acres from 0.25 acre). Even in this relatively open environment, the animals are restricted in comparison to the wild environment. Surely examination is needed of changes and consistencies as a function of environmental context in a well-studied group of individual animals. It is clear, however, that patterns of behavior observed under any given circumstance reflect, albeit indirectly, evolutionary pressures which have operated over countless generations. When judiciously applied, studies of wild animals in captivity can serve a critical role in the documentation of boundaries of phenotypic adaptation in behavior—a question of obvious critical importance. Such studies, however, particularly with reference to the dynamics of social relationships, are at a very preliminary stage.

It is the question of boundaries of the organism's behavioral adaptations to environment that we eventually wish to explain. Our techniques are not yet sufficiently precise or sensitive to give us definitive answers. Yet it is obvious that progress can be and has been made. Further pursuit of these questions can help us gain fuller appreciation for the animals with whom we share this planet, as well as generate a more satisfactory evaluation of the mechanisms which underlie their behavior.

EPILOGUE

Many insights that we may gain in the future are likely to derive from our attempts to deal more adequately with subtleties of relationship and change in the social behavior of these animals. One possible, considerable, and relatively unexplored potential of captive studies stems from detailed monitoring of the relationships between animals and their human handlers. We have been struck, for example, that certain individual handlers can very quickly establish a comfortable working relationship with these animals, while other handlers fail to do so. We are reminded that Kramer's wolf (Kramer 1963) attacked him when it was ten months old, while we have never been seriously threatened during the dozen years over which our hand-rearing and related studies have been conducted (cf. Fentress 1967).

We have begun to document the interactions between animals and people to help us clarify some of the rules that may be employed in social relationships between the animals. The great advantage of this approach is that we can analyze "mistakes" as well as "successes" in human-animal interactions and then compare these data with that obtained for interactions between the animals in various social contexts.

For example, recently obtained video tapes suggest that human movements which appear too slow and controlled (and, in retrospect, stealthy) can elicit avoidance just as readily as movements which are too rapid. There are times to withdraw one's gaze from a given animal and times to entertain oneself—such as by playing with a bit of grass. There are times to reduce avoidance tendencies by initiating active "games" rather than by remaining still; and there are times to withdraw momentarily. The success or failure of these movements can be compared to data obtained from observed interactions between the animals. Such observations can provide important insights about the cues and modes of behavioral expression used by animals themselves.

It is striking to note how much easier it often is for a skilled observer to mimic various postures that seem appropriate for a given social context than to verbalize these postures initially. Rather parodoxically, it is often those who can verbalize *a section of* the total complex of postures and cues used in a particular situation who have the most difficulty performing correctly within a broader, initially more intuitive sense. Through appropriate documentation procedures, however, these more intuitive complexes of relationship can be translated into formal terms.

It is important that the human observer not only seek to pick up fundamental cues from the animal, but to translate his/her own behavior into cues that can be understood by the animal. For example, one of us (Parr) found that upon entering a den to film an adult female and her newborn cubs, the female tensed her muscles slightly and made "intention" movements to leave her pups—until the human observer closed her eyes, nodded her head, and sighed deeply (as often observed in "relaxed" social groups of wolves). This behavior was sufficient to relax the female so that successful films and recordings could be made.

We suspect that it may be equally important to provide the animal with an opportunity to display toward a human observer certain forms of behavior normally observed in encounters between animals. For example, by extending one's arm toward the face of ambivalent animals, they may tug on it rather like they normally tug at the faces and tails of each other. A likely, not yet fully substantiated consequence of allowing the animal to perform this species-characteristic behavior is that social relationships between animal and human are facilitated. This observation also provides important clues to the human observer about the classes of behavior she or he should perform.

Often when wolf pups play with one another or explore objects in their environment, they paw other animals or objects repeatedly. Even with quite timid animals it is possible to elicit this pawing by moving one's hands just in front of, or beneath, the animal's forefeet. Once the pawing is initiated, the interaction may be intensified in a manner previously impossible. Such examples could be extended indefinitely. What is important to realize is that when properly documented, such observations can provide very valuable, relatively untapped insights into the organizational principles of social communication and behavior.

We can benefit as much from our "mistakes" as from our "successes." One of us (Fentress) frequently participated in howling sessions with the adult male (Lupey) from outside the animal's pen. These howling sessions were marked with much tail-wagging and generally "friendly" behavior on the part of the wolf. On three separate occasions, however, the same observer attempted to initiate howling from *within* the wolf's pen. In each case, this attempt elicited snarls and bristling of fur by the otherwise extremely gregarious wolf. This example clearly indicates that the animal's response to the same stimulus (human howls) varied as a function of the context (inside versus outside the pen) in which the stimulus was presented. It is not difficult to generate suggestions about the diverse functions of howling in the field which, in turn, could be tested empirically.

Our level of discourse here is quite different than in previous sections of this chapter, but it may provide an important complement toward understanding these animals in greater depth. The research worker must walk a very delicate line between distant formalism and more intuitively based judgments if he or she is to interact successfully with captive wild animals in addition to studying them in the usual sense. These intuitively based judgments, however, complement rather than contradict more formal analyses, and it would be a tragedy of modern science if they were considered off-limits.

But there can also be a more profound error should these subtle routes of information be excluded from our consideration: the study of wild animals in captivity demands an appreciation and ethical concern for the creatures with whom we work. If their welfare depends in part upon our ability to read the subtleties of their behavior, then further pursuit of these subtleties is a major obligation. And we believe that such efforts will pay off in objective knowledge as well.

4 Experimental Analysis and Control of Group Behavior

Hal Markowitz and Gail Woodworth*

> Undoubtedly, kindness to captive primates demands ample
> provision for amusements and entertainment as well as for
> exercise. If the captive cannot be given the opportunity to work
> for its living, it should at least have abundant chance to exercise
> its reactive ingenuity and love of playing with things.
>
> The greatest possibility of improvement in our provision
> for captive primates lies in the invention and installation of
> apparatus which can be used for play or work (Yerkes 1925).

THE NATURAL OR UNNATURAL ENVIRONMENT

When a human being becomes hypertensive or obese, we
expect that the prescription for remedy will include loss of weight
and a regulated program of exercise. When we become exasperated
with something in our everyday lives and do not have the physical
outlets which some of our ancestors may have had to utilize the
effects of glandular secretions, we sometimes develop ulcers or
other physical disabilities. Again, the common prescription deals
with finding ways to expend this energy, often in a highly artificial
fashion (e.g., rowing devices that take you nowhere or bicycles that
stand in place). Why, then, does it seem so unthinkable that an
animal in captivity might need similar artificial methods to cope

*Hal Markowitz is Director of the Oregon Zoological Research Center, Portland,
Oregon. Gail Woodworth is in the Graduate Psychology Program at Portland
State University, Portland, Oregon.

with the restrictive environment? Why does the nagging feeling persist that the pushing of keys, pulling of chains, stepping on treadles, and similar activities are normal behaviors for humans but unnatural for other animals? The variety of social behaviors documented by ethologists (e.g., Bermant and Alcock 1973; Eibl-Eibesfeldt 1970; Hess 1962; Hinde 1959; Lockard 1971; Lorenz 1970; Morris 1970; Tinbergen 1953; van Lawick-Goodall 1968) is exciting and even astonishing to those who think of animals as "simple" or even automatic.

Captive animals can adapt to forms of food which none of their wild counterparts might encounter, providing that such food contains sufficient nutritional value. It is untrue that animals are necessarily more "instinctive" in their behaviors than humans, or that they are naturally retarded in the possible variety of their ventures (Beach 1955; Lehrman 1953). We suggest that many of our ideas about the stereotypic behavior of animals derive from environments which by design or omission are so impoverished that they allow minimal stimulation. Complex animals develop their own idiosyncratic ways of responding to a particular milieu much as we do. If we take the time to teach them, they can learn seemingly miraculous routines (e.g., Breland and Breland 1966; Kellogg and Kellogg 1933).

Independent of terminology, it is clear that the environment plays a large role in determining every animal's behavior. It is time to purge ourselves of influences which demand that we focus on some natural behavior without reference to the environment. In the same sense that we would want the richest possible experiences for other human beings within their milieu, we might strive for similar goals in designing environments for animals.

ANTAGONISM TOWARD BEHAVIORAL ENGINEERING

The opening quotation from Yerkes makes clear that the notion of instrumentation with which to evaluate animals' behavior and provide them healthful activity is hardly new. But a paucity of experimentation in this area exists. Reasons for this lack are manifold. First, selection of articles by journal editors is usually based upon long-established norms related to arbitrary confidence levels. Indeed, the "scientific method" so heavily instilled in most of us demanded that research data be derived in situations where as many potential outside "variables" as possible were "controlled" in order not to interfere with our focus on "independent variables." It is no wonder that the vast majority of hard-core scientific "truths" in animal behavior have been derived from study of individual organisms virtually *in vacuo*. Some of the specific observations detailed in later parts of this chapter suggest that many behavioral "laws" may apply *only* to solitary animals. Adequate taxonomies of group behavior are notoriously difficult to establish (Altmann 1965; Bobbitt et al. 1969; Box and Poor 1974; Chamove 1974; Hinde and Atkinson 1970; Hopf 1972).

A second major problem concerns the apparent cost of the innovations to be discussed. We have shown (Markowitz 1975) that this cost need not be prohibitive and, in some cases, may be self-liquidating. In any case, humane care should include budget considerations which provide a less than barren behavioral environment for our captive animals.

As earlier suggested, a third reaction arises from the "naturists" who are against the introduction of group-behavioral apparatus within institutions such as zoos and wildlife stations. We have noted elsewhere (Markowitz 1973, 1974) that behavior considered natural for an animal in captivity is hardly comparable with what might be expected in the wild. Should we, then, expect that observed behaviors of bears in grottos, monkeys in wire cages, and hoof stock in pens are specific to that species—or are they determined by the environment in which the animals are housed? It is always necessary to remember that in the wild, as opposed to the typical zoo environment, most species exercise a significant amount of control over their own lives. Certainly, if we are to study exotic animals and hold them institutionally for our education and entertainment, we owe them every reasonable provision which may enhance their lives.

BEHAVIORAL ENGINEERING STUDIES AT THE PORTLAND ZOO

Due to the uniqueness of the zoo setting, carefully controlled research is extremely difficult. Despite evidence that we must expand beyond situations where traditional and simplistic experimental procedures may dictate results (Kendler 1965; Restle 1965), it is quite unsettling to be faced with the real problems of a complex zoo society. Interaction with personnel who believed that old ways were the only ways quickly identified us as intruders in the eyes of many. In the absence of well-established precedents, much of our work was accomplished on an empirical basis, with each new step in behavioral engineering dependent on the outcome of immediately previous procedures. As the project commenced, we decided to collect every kind of data that our limited budget would allow. For example, a print-out counter was used to collect information about all responding in one-minute bins, twelve hours per day for the first year. Even with computer assistance, data reduction is an awesome task. Multiply this requirement by the time factor involved—that our "experiments" are usually measured in months and years rather than in some short-term criterion—and one may appreciate why results of these behavioral engineering tasks can here be presented only in summary form.

GIBBONS

In the wild, gibbons run in beautiful patterns through trees, collecting their food and sometimes passing over many easy morsels for apparent enjoyment of the exercise (Carpenter 1964; Ellefson 1967; Tuttle 1972). Captive gibbons may occasionally put on spectacular displays and are especially responsive to crowds. In the Portland Zoo, one of our animals, Harvey Wallbanger, entertained visitors by doing triple bar flips, flying thirty feet through the air, and barely avoiding obstacles before catching another portion of his exercise apparatus. Unfortunately, his cage mates rarely did more than watch Harvey during his routine. We wanted to find a way to encourage them to be active too, since Harvey appeared by far the healthiest, happiest animal. As in most zoos, our gibbons were fed by having their daily ration thrown to them on a cement floor. Watching these arboreal animals sit in awkward positions and pick over piles of food seemed not only uncomfortable but demeaning, and a poor educational experience for zoo visitors as well.

Without a large budget it was impossible to provide a forest for gibbons with appropriate climatic conditions. The first of our major behaviorally engineered exhibits demonstrated, however, that it was possible to encourage brachiation and "flying" around

the cage while simultaneously providing the animals some entertainment and control over their own feeding.

The gibbon (*Hylobates lar*) colony at the Portland Zoo
included two mature males, one female, and an infant male when
research began. This exhibit represented an unusual challenge
because we were dealing with a mixed group which was never
removed from its home cage.

Prior to installation and design of the equipment,
considerable time was spent in consultation with primate-house
staff about convenient locations which would reduce impediments
to routine care (Markowitz 1975). The home cage was a relatively
large enclosure equipped with two openings, into each of which a
special panel was mounted, about thirty feet apart midway up the
back wall (approximately fifteen feet above the ground). Each
panel had a large lever and stimulus globe. The right-hand
apparatus also had a food dispenser. During pre-experimental
observation there was some sharing between the animals, but no
significant dominance hierarchy was apparent.

Initial shaping was accomplished by requiring successive
approximations to pressing the right-hand lever. Reinforcement
was delivered automatically, which allowed the experimenter to be
in front of the cage, and in a few days all three full-grown cage
residents were successfully operating the manipulandum to obtain
pieces of apple, orange, banana, carrot, and monkey chow.

The second portion in our goal of establishing a behavioral
chain was much more demanding. In order to shape the animals to
both stimulus panels, we began by delivering reinforcement for
movements away from the payoff station. There were predictions
that we might never succeed or take years to establish the behavior;
but our gibbons were such adept students that in approximately
one month, each was able to complete the entire sequence (see
Figure 4.1). By the tenth week of our work, they were averaging
more than 114 round trips per day, thus significantly increasing
cage activity and simultaneously providing zoo visitors with a
demonstration of their learning ability.

After several months, the public was invited to initiate trials by
depositing coins into a box in the viewing area. Considerable care
was taken in the design of the accompanying instructions:

Figure 4.1 Gibbons earning their own food by movement between platforms.

Research Contribution:
Ten cents will start a trial when the light on this box is lit.
The counter shows the total number of pieces of food earned by
the gibbons today.
Animals are not machines and the gibbons may choose not
to respond when the light is turned on. All money collected
here will be used to develop more activities for our animals.

The only "complaint" from a zoo visitor arose when the
apparatus was being serviced and he was unable to put a dime in
the machine (which he had returned to the zoo especially to do).
Generally, people have been excited about making a contribution
in this form. The thousands of dollars in dimes which this
apparatus yearly collects significantly help to produce additional
new exhibits. The apparatus is continuously available to the
gibbons from 10:00 A.M. to 10:00 P.M., and if no zoo visitor initiates
a trial, a timer automatically turns on the apparatus two minutes
after each piece of food is earned. These animals have potential
access to food for a greater portion of the day than have the rest of
the primate collection. We were fortunate to observe informally
some indications of the animals' "feelings" about the apparatus. A
change from electromechanical to solid state equipment resulted in
a considerable period when the apparatus was inoperative. Since
the gibbons had been free to feed at any time of day, we continued
free food access during this period.

The gibbons during this time spent a considerable amount of
time (even when holding food in one hand) trying to turn on the
apparatus. Such attempts included mounting and hugging the
lights and apparently "imploring" the apparatus to respond.

The fact that more than one animal had simultaneous access
to the apparatus led to some enticing questions about social
interactions. Once shaping was accomplished with all the animals
in a single large cage, would they steal food from one another?
Might they develop some cooperative methods of reducing the
work requirement? The answers were bound to be interesting since
the literature was devoid of similar "social" behavioral engineering with captive animals. Despite our readiness for unique data,
the first three years of this work revealed some behaviors which we
had to see repeated many times before we were willing to believe

them. The two mature males we began with were different from
one another not only in their social behavior, but in their responses
to this new challenge.

Harvey Wallbanger (so named. because of his characteristic
wall-banging in mid-flight) eventually began to do almost all of
the hard work. Kahlil, Momma, and Super Squirt (who was still
nursing at the beginning of our work) were all willing to expend
the effort to pull the lever which led to reinforcement. Predictably,
Harvey would tire of being the family provider and would not
respond to the remote stimulus until the others backed away from
the payoff lever. A spectacular set of races would ensue, in which
Harvey became so adept that he could accomplish the whole
required sequence in less than two seconds. The gibbons "worked
out the details" so cleverly that a dead heat often resulted, and it
was a matter of chance which animal got the food. For a period of
approximately one year, Harvey relented with respect to his mother
(Momma) and would feed her without hesitation. Whenever Kahlil
went to the payoff station, however, Harvey would quit work until
Kahlil backed off sufficiently for a contest.

As the infant matured, he learned the complete required
sequence, although he was never individually shaped to perform
the task. Much of this gibbon's behavior was apparently learned by
observation. Three years after inception of the experiment, Super
Squirt has become a spectacular swinger and does at least as much
work as Harvey. The gibbons, who have now earned all of their
food for more than four years, are providing a very popular public
exhibit and are consuming slightly more food than prior to
behavioral engineering. It is noteworthy that remarkable lack of
scrapping occurs over food and that absence of old food on the
floor improves cage appearance and makes the keeper's job somewhat easier.

DIANAS

As is usually the case with intelligent and complex subjects,
our work with gibbons produced more questions than answers.
Principal questions concerned cooperation, "rip-offs," and altruis-
tic behavior. The most experienced primate observer would have
difficulty deciding whether, when two animals arrived at the

payoff lever simultaneously and broke a piece of food between them, the behavior represented sharing or stealing. To further address these questions, we established a token economy with a family group of diana monkeys (*Cercopithecus diana*). At the onset of the experiment, the group consisted of a sixteen-year-old female (Beulah), her eight-year-old mate (Rocky), their adolescent (Butch), and infant (Kid) offspring. During the course of the work, two additional offspring were born.

The subjects were initially shaped, through successive approximations, to exchange poker chips for pieces of fruit or monkey chow. This was accomplished by having a monkey deposit a chip into a V-shaped funnel which channeled into a coin slot. Much to our surprise, every animal except the adult female learned the desired response. Beulah's best approximation after months of training was to sit on the feeding station with a chip in her hands and drop it on the cage floor.

Rocky would consistently share his pieces of food (banana, orange, apple, monkey chow, or carrot) with Beulah at the feeding station, but defended his earnings from the rest of the family. Luckily, the juvenile quickly learned the task and became more willing to distribute the products of his efforts.

We finally decided that we would progress to the next stage of shaping despite Beulah's apparent lack of success, since she was getting sufficient food. From this point on, the feeding station was *always* available for token acceptance. The dianas were shaped to accomplish a sequence of behaviors similar to that described for the gibbons, but with a payoff of tokens rather than direct food delivery. They were first taught to pull a chain at the token delivery station which was approximately fifteen feet above, and twenty feet to the side of, the feeding platform. Final shaping trained the monkeys to move all the way to the top of the cage and pull a long chain which activated the token delivery station (see Figure 4.2).

Because the adolescent and the infant began to swing on the upper chain almost immediately following its installation, this final step was accomplished with ease for all subjects except Beulah. Eventually, Beulah settled down to a specific set of strategies, encouraging the others to work and occasionally pulling one of the chains where the token was delivered. Yet she always

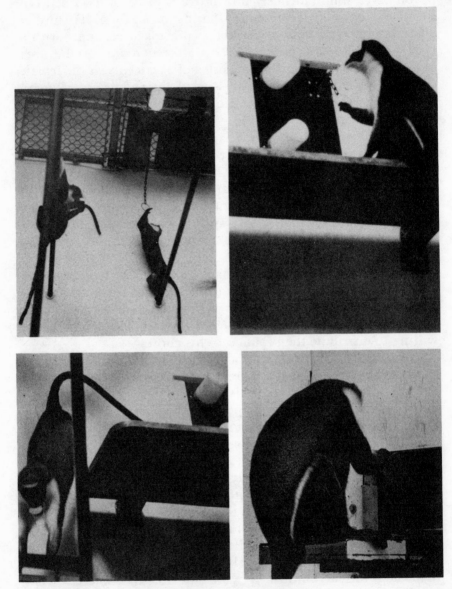

Figure 4.2 Diana monkeys' token economy behavior.

gave the token to some other animal, from whom she would "rip off" the acquired food whenever possible.

Many unexpected systematic "secondary" results have been observed. For example, Rocky showed remarkable stereotypy for two years in accomplishing the sequence. At the top flight of the cage, which included paths in almost every direction, he would invariably walk past the bar leading to the chain at least once and *always* make a right turn as he approached the first response requirement. The youngsters showed much greater versatility, taking all sorts of shortcuts depending on their starting position. Initially, the adult male dominated the scene, taking over the food-earning sequence whenever he chose, and the youngsters waited until he was otherwise occupied to earn and exchange tokens. As time progressed, however, the dianas would often pull chains for one another, sometimes sitting quietly watching the other animal respond at the payoff station.

Although we continue supplementary feeding to be certain that none of the younger animals are deprived, we see "softening" of the male's behavior as he becomes more and more adept at "breadwinning"—and we see occasions when he gives the youngsters tokens or food during the experimental session.

Behavior has stabilized substantially during the last year as shown in Table 4.1. Besides basic data collected on this token economy, activity generated by the apparatus has provided opportunity for a number of observational studies by other researchers (e.g., Soper 1974; Bandura 1974). Watching zoo visitors become "cheerleaders" and gain respect for the agility of diana monkeys is another fringe benefit (Chasen 1974).

Table 4.1 Mean Number of Tokens Dianas Earned Per Day

Number Working Days	Mean Number Tokens
90	124.39*
90	130.07
90	135.63
90	130.55

*Shaping included in this session.

HARBOR SEALS

The first harbor seal studies conducted at the Portland Zoo used a chamber designed to isolate a single seal. Three of the four subjects were trained to move out of the water into this enclosure (see Figure 4.3).

The seals first pressed panels to obtain smelt. Then a standard light-dark reversal discrimination was begun. Mechanics of this procedure were as follows: the side upon which the light appeared was randomized, and the animal's first problem was to select the place where the light appeared; after that was accomplished to a criterion of twenty successive correct responses, the situation was reversed and dark was correct.

As time passed, we became increasingly distressed that we were not abiding by some of our own suggestions about viewing animals in a more complex milieu (Markowitz 1974). In addition, we frankly had a lot more fun watching the incidental learning of our seals when they swam freely and interacted with us as we waded among them. So we decided to build a new apparatus which would allow the animals to swim freely and, at the same time, allow us to observe their learning abilities and effects of the social situation (see Figure 4.4).

The new apparatus consisted of two twelve-inch stimulus globes spaced about twenty-four feet apart. The globes were located approximately one-and-a-half feet from the edge of the island contours and three inches from the waterline.

Our seals Milhouse, Neptune, and Olin devised the most elegant gamesmanship and did everything imaginable to complicate our data. Summary results are shown in Table 4.2, but tabular presentation masks the richness of the behavior represented. We will try to give a brief sample below.

During the first stage of training in this free-swimming situation, we expected that some transfer from the light-dark problem previously learned by each animal in the single chamber might occur. Instead of our "educated" seal Milhouse initially responding, however, it was Neptune, heretofore only able to watch the other seals working through the glass front of the chamber, who began to respond on the first day. He solved the initial problem to a criterion of twenty consecutive correct in 696

Figure 4.3 Seal in individual testing apparatus.

Figure 4.4 Seals in free-swimming test situation.

Table 4.2 PERFORMANCE FOR HARBOR SEALS ON REVERSAL DISCRIMINATION

Days	Reversals	Trials to Reversal	Percent Correct Responses
60	3	12,929	62.2
60	3	13,437	68.0
60	13	15,973	75.3
60	44	10,524	77.5

trials. When the situation was reversed, however, he quit after several abortive attempts, and the task was taken over by Milhouse. During all of this early work, our procedure was to deliver reinforcement on the side where correct responses were made approximately twenty feet from the manipulandum. This resulted in some spectacular races and decoy behavior, with the animal who made the correct response racing with the "poacher." As trials progressed, two of the seals learned so well to watch the behavior of the responding animal that they beat him to the fish most of the time. Therefore, we eventually changed the procedure. Beginning with the fifth reversal, reinforcers were delivered to one of three locations according to a random series. This meant that the fish became quite uniformly distributed and that much of the elusive behavior so interesting to us was now eliminated.

Our next startling result involved a herring gull which began to visit each day at 10:30 A.M. and 2:30 P.M., when we ran our sessions. The bird would stand high on the roof and watch with apparent interest as the seals worked and the fish were thrown to them. Suddenly during one session, it swooped down at precisely the right time and made off with one of the smelt, just as Milhouse was swimming toward the payoff. The resultant behavior was quite remarkable: Milhouse raised much of his body from the water and moved his head from side to side. He stopped responding and moved onto the island, refusing to work. Neptune, who had solved the first problem and had not participated for fifteen weeks, "took over" and did much of the work for the rest of the session.

We have described this anecdotal set of observations in detail because it outlines one of our major suggestions. In the laboratory, one would most likely abandon the research, saying that uncontrolled variables made the results unusable. But in our circumstances, we suggest that such complications make the results even richer and more interesting. The laws of learning in a situation containing even minimal complexity are certainly not as straightforward as those demanded by traditional criteria. They may, however, be more important in the "real world"; and also, infinitely more interesting.

GIBBONS: REVERSAL DISCRIMINATION

A new pair of young gibbons (a four-year-old male named Milo and a seven-year-old female, Venus) was added to our collection and housed in a large display separate from our existing colony. An apparatus with press panels was installed, and a standard light-dark reversal discrimination procedure, as with the harbor seals, was used with food delivered automatically for correct responses. This work differs significantly from similar testing described by Essock and Rumbaugh in chapter 6 of this volume because both gibbons had access to the apparatus throughout the sessions. When an individual organism is tested on this sort of task, the initial discrimination is relatively quickly learned by a naive animal. But numerous trials are required to reach the first criterion on reversal. Repeated reversals, depending upon the particular animal and species (see Essock and Rumbaugh, chapter 6), eventually result in a reduced number of trials to criterion, instead of an accelerating number (e.g., Markowitz and Becker 1969).

Initially, we had planned to require that both animals accomplish the criterion (twenty successive correct) before each reversal. We found that Milo learned the light-correct problem very quickly, while Venus took considerably longer. When reversed, Milo made no apparent progress toward learning dark-correct, and Venus "took over" the new task. After considerable contemplation, we decided that it would be interesting to run a cage criterion (i.e., as long as total responses for that cage accumulated twenty in a row, it didn't matter which animal did the pressing).

Table 4.3 shows that the gibbons have become somewhat specialized. Milo makes light-correct responses with virtually no errors, while Venus has done much better on dark-correct. Persistence of this "specialization" (through 1,339 reversals) was unexpected and provides an interesting avenue for analysis of differences between group and individual performance.

Table 4.3 MEAN NUMBER OF CORRECT RESPONSES AND AMOUNT OF PARTICIPATION FOR LIGHT-DARK REVERSAL DISCRIMINATION OF GIBBONS

Total Days	Performance	Light Correct		Dark Correct	
		Venus	Milo	Venus	Milo
90	Total Responses	22.75	18.75	97.82	11.62
	Correct Responses	17.78	18.74	93.76	.00
90	Total Responses	27.61	11.20	64.87	6.84
	Correct Responses	23.89	11.19	60.46	.03
90	Total Responses	21.74	21.56	165.27	22.94
	Correct Responses	16.76	21.47	160.15	.27
90	Total Responses	15.04	20.79	110.92	15.17
	Correct Responses	12.33	20.42	107.55	.17
90	Total Responses	19.80	11.84	65.63	8.95
	Correct Responses	16.91	11.84	62.80	.02
90	Total Responses	19.95	11.29	45.48	7.80
	Correct Responses	17.71	11.15	43.55	.39

After nineteen months of running two 1-hour experimental sessions daily, we have begun a totally automatic procedure which allows the gibbons to work whenever they wish from 10:00 A.M. to 10:00 P.M.; and we are now seeing an average of ten *reversals* every day. Analysis of these extensive data will provide information not available in the current literature about complex habitual tasks. We have intentionally avoided individual shaping to improve

Milo's responding on dark-correct or Venus's performance with light in order to evaluate the persistence of these behaviors.

SERVALS

A discussion of preliminary work for our first engineered exhibit with felines may help communicate the extensive preparation demanded by each new regimen. Almost a year's work was invested before construction of the automation was begun.

In the wild, the serval (*Felis serval*) is noted for the agility and quickness with which it captures prey (Ewer 1968). When the potential meal is a game bird, the serval may flush it from the brush and catch the bird in flight with one leap. As suggested by Cheney in this volume (chapter 1), most of the postural and graceful beauty of these animals is lost in captive environments where food is simply thrown on the floor. After an extensive baseline study (Schmuckal 1974) which identified periods of both excessive pacing and inactivity, we began a project designed to recapture some of the active behavioral beauty and to provide some entertainment for the servals. An apparatus which "flies" a piece of meat in random patterns and heights across the serval exhibit is under current construction. When the serval leaps and "captures" this prey, it has a quick snack coupled with the exercise so often missing in captivity. We were honestly unprepared for the results of our preliminary study to calibrate height requirements for the target (see Figure 4.5). Just as we were starting to reel down a meatball on the line, the animals perked up, looked twice as tall as they ever had before, and leaped two body-lengths straight up in capturing the food.

Schmuckal's baseline study established that these animals were quite dormant and stereotyped in their behavior prior to initiation of new feeding procedures. The study also provided other significant data pertinent to questions about social interaction. We were concerned about the possibility of aggressive behavior when food was distributed in the relatively restrictive environment. Since we had three servals, our first approach was to drop three spatially separated pieces of food at a time, and to observe the effects not only on food consumption but on other behaviors as well.

Figure 4.5 Serval leaping for food in initial trials.

In general, the servals' behavior became quite predictable. The male took whatever he wanted, usually at least two out of three pieces of food, early in the session. One of the females became the second-order dominant animal, while the third animal was always the last to begin obtaining food. There was no significant aggression, however; and, since servals tend to be skimpy eaters anyway, it was seldom very long before the last female was sharing food.

CRITICAL PROBLEMS

A critical look at behavioral engineering in the zoo thus far suggests the need for even more extensive baselines than are typically established. With groups of animals in limited enclosures, every hint of potential sources of aggression or maladaptive routines which might be generated by the equipment should be identified.

We think that one imperative step in behavioral engineering for groups of animals is the adequate assessment of possible deleterious effects, on other behaviors, of instituting ways for the animals to gain some control. Interestingly, in those species where increased aggression has occurred as a function of apparatus introduction, a consistent pattern is obvious as long as the stimulus situation remains novel. Some transient increase in aggression may occur, but once the animals become acclimated to their new opportunities, we typically see ratings of aggression diminish below the baseline levels established prior to our work.

Conferring with other colleagues who have extensively studied behavior in some of the same species, we would tentatively propose the following hypothesis: aggression is directly related to novelty when obtaining food is at stake. With animals that display more aggression with increasing environmental complexity, the effect diminishes with familiarity (see Figure 4.6).

In most cases, as illustrated by our discussion of the serval project, important intermediate steps should be taken prior to final exhibit design. Much equipment alteration may be avoided by manually testing individual portions of the routine prior to automation. An important reason for this step is the great difficulty in estimating the skill and physical strength with which

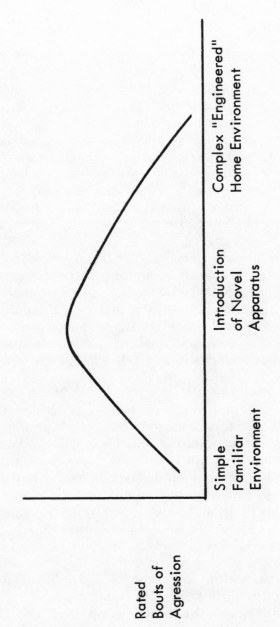

Figure 4.6 Hypothetical curve relating aggression to situational novelty.

exotic animals will attack new tasks, as shown by the servals who surprised us with the extent of their vertical leaping.

It has also been apparent in our work that apparatus should always allow for additional tasks to keep life interesting for the animals. Now that zoo visitors have learned that our primates are capable of much more intelligent behavior than making faces and exhibiting stereotypy, we researchers are slowly learning that apparently complex tasks may become "child's play" for diana monkeys in a matter of months. In future exhibits, we hope to allow multiple activities leading to reinforcement so that the resident animals can select in cafeteria style the activity in which they wish to engage.

In retrospect, this chapter might have dealt more extensively with the importance of studying species' typical behavior. But we have presumed that most readers will understand the importance of allowing animals to perform tasks for which they are well-adapted. No better initial suggestion can be offered to a person interested in beginning behavioral engineering than this: study the behavior of the selected species in as much detail as you possibly can.

What Has "Gone Right"?

Using multiple subjects has enriched rather than impoverished our data. As suggested at the beginning of this chapter, analysis of the resulting complexity is awesome task. But, in a real sense, even this complexity may represent a benefit. We can, much as with ethological observations (van Lawick-Goodall 1968), allow an almost infinite number of "piggyback" studies for other researchers. Establishing opportunities for animals to arrive at their own conventions for distributing work and earnings introduces to an otherwise barren social milieu many of the most interesting aspects of social behavior in the wild.

The health care of those zoo animals involved in daily experimental routines receives special attention, not because other zoo residents are overlooked but because daily observation periods will inevitably tend to identify gross physical abnormalities. In some cases, such as our elephant memory studies reported elsewhere (Markowitz et al. 1975), we have identified health problems (in this case, an apparent pronounced deficiency in retinal vascularization in two of our elephants) which probably would never have been discovered in the ordinarily monotonous daily routine. Our veterinarian is fond of saying that the chimpanzees now learning sign language in our zoo receive better maternal attention from volunteer surrogate mothers than any child he knows. When the chimps sniffle, the doctor is called without hesitation.

We think that the paramount criterion in behavioral engineering should always be consideration of the benefit that new apparatus will provide for the resident animal. Our exhibits are intentionally *unlike* carnival exhibits. Long periods occur during the day when the animals select to ignore the apparatus and engage in other social activities.

Instead of designing a schedule that compels the animal to respond every time a meatball flies past or a token becomes available, we emphasize offering the animals opportunities to do a little healthy work whenever they choose to feed themselves.

5 Applying Behavioral Knowledge to the Display of Captive Animals

William A. Myers*

EMERGING FROM RESEARCH: A BABOON EXHIBIT

Creating a behavioral display in which a baboon could compete with a human in a reaction-time task was, it seemed, just a diversion. I never doubted my prompt return to the series of experiments I had been conducting for two years on the phenomenon of observational learning (or imitation learning, an older, less precise term for the same phenomenon).

In the experiment which gave rise to the reaction-time contest, a four-year-old male black baboon had already learned to push any lighted button in a column of three. The next step would have been to record how long it took the observers (in this case, three juvenile rhesus monkeys) to match the performance level of the demonstrator animal.

Instead of taking that next step, however, I made a second panel of push buttons so that a human could compete with the baboon in a reaction-time contest. One of the three lights on each contestant's panel would come on simultaneously. Whoever turned it off first scored a point. So that points were clearly visible to both contestants, they were recorded on a column of six colored pilot

---

*William Myers is a behavioral engineer specializing in the application of behavior principles in zoos and aquaria. His particular interests are displays of natural behavior and the design of captive habitats.

lights. A player's first point would light the bottom lamp in his score column. The next point would light the lamp second from bottom, and so on. When the topmost light on his score column was lighted, a player's next point would activate an automatic feeder that dispensed a food pellet. These 190-milligram whole-diet pellets were the primary reinforcers for the animal. Humans rarely ate the pellets they won; the baboon always ate his. And the baboon won the great majority of contests (about 87 percent for duration of the exhibit).

When the baboon's rate of turning off lighted buttons in less than a second had reached 85 percent, I sat down to compete with him. The first contest was a breeze for the animal. And the second. On the third trial, I finally beat him. When he heard the pellet drop into *my* tray, and when he saw *me* pick it up and eat it, I was treated to a display of emotional behavior that few laboratory workers get to see. Teeth exposed, lips curled back, hair on the back of his neck standing up, he gripped the mesh of his cage with all twenty digits and proceeded to shake that cage so hard that I involuntarily left my seat and retreated to the door of the room. The accompanying shrieks were the loudest I ever heard. Except for very subtle, attenuated displays of emotion, that was the one and only time he ever reacted to losing a game.

The behavior of my noontime visitors, who lined up outside the lab, was as interesting as the baboon's. Nearly everyone sat down to play with a smile. When they lost the first game, the smile faded. When they lost the second game, they began to look very determined, despite calls of "Hey, it's my turn." If they won the third game, they exited smiling a different kind of smile, embarrassed to have demonstrated how important it was to them to beat that animal.

Humans need only a game or two to play competitively. But the baboon acquired his skill through a series of discrete training steps.

The first step in training the animal was simply to expose him to the buttons. For the first thirty minutes, he kept picking at the edge of the buttons with his fingernails. By the end of an hour, this behavior was infrequent enough to let me begin the next step, which was to teach the animal to push the button with enough force to activate a microswitch. At 95 percent of his free-feeding weight, and with that hour of exposure to the buttons behind him, the baboon required only another hour to acquire the new response of extending his forefinger and pushing the buttons.

In the next session, he learned to respond only to lighted buttons. In the last twenty minutes of this two-hour session, 90 percent of his responses were to lighted buttons and only 10 percent to unlighted ones. Then the number of correct responses required to produce a pellet was gradually increased. Within an hour, he stabilized on a ratio of three responses per reward (FR3).

Concurrent with the increased response requirement, the circuit was wired to impose a decreasing time period in which to make the response (limited hold). To begin with, the animal had 10 seconds to respond after the button lighted. That time was reduced in decrements of 0.5 second until the animal was responding in less than one second 85 percent of the time.

A fixed time period of four seconds existed between the appearance of one lighted button and the next. Within two days of employing this four-second intertrial interval, the animal had acquired recognizable patterns of behavior to fill the time between trials. Either he licked the top of the response panel a few times or he got up and turned around. In either case, he was seated before the response panel in time for the next trial. These superstitious actions were as much a part of the animal's chain of behaviors as the ones for which he was deliberately reinforced.

Adding the score panel came next. In all, seven correct responses were required to earn a reinforcer. The first six produced a conditioned reinforcer in the form of a change on the animal's score panel. The seventh delivered the unconditioned reinforcer and turned off all score lights. The schedule of reinforcement at this point was as follows: using a discrete trials procedure with a constant intertrial interval of four seconds, completion of an FR7 with added counter (the score panel) produced a reinforcer.

At the end of two hours under this procedure, the baboon was punching 87 percent of all lighted buttons in less than 0.75 second and was punching unlighted buttons at an average rate of 1.3 per minute.

With those six hours of training completed, I set up a second identical panel of buttons and a second column of score lights and began competing with the animal, as previously described. At this point, every correct response by his human opponent postponed reward to the animal, making the schedule a variable ratio. What's more, whenever a human won a game, the baboon's added counter would reset to zero without his receiving any reward.

But the animal's skill at the game won him such a high percentage of games that the reinforcement schedule over a day's play resulted in a figure close to the fixed ratio that prevailed when he played the game alone.

THE JAGUARS WERE SISTERS

The director of the Como Zoo in St. Paul, Minnesota, understood that this type of applied behavioral work could help solve some practical problems in his zoo. He had two young, hand-reared jaguars that spent much time in aimless pacing. Was there a way, he wondered, to change that behavior?

The most straightforward way to reduce the frequency of pacing, I thought, was to teach the cats a new behavior that would be incompatible with pacing.

This new behavior might just as well provide the public with something to see and display some quasi-natural motor movements in jaguars. These two goals could be met by installing a device which the cats would have to strike with their forepaws while standing on their hind legs. Rearing up and striking with a forepaw are well-practiced movements in feral cats, and even zoo cats use these movements from kittenhood on. The new part for the cats was to emit these behaviors toward a novel object, which bore very little resemblance to the kinds of natural objects that might elicit rearing and striking in the wild.

Before the project got started, some of the lore about jaguars started drifting my way. A seasoned veteran of zoo and circus told me that jaguars would not work for "bait" (i.e., food reinforcers). Jaguars share with other big cats a feeding pattern in which large amounts of food are eaten at widely spaced intervals. Would such an animal work for small pieces of food?

If not, the obvious move would be to gradually increase food deprivation until smaller pieces of meat became attractive as reinforcers. That was not possible, however, because I had agreed to work within the daily feeding schedule set by the zoo. Clearly, the only way to get going was to attempt training the animals under eighteen or twenty hours of deprivation and be guided by the outcome.

An automatic feeder holding fifty-five pieces of horsemeat was mounted and serviced from the keeper's alley. The cats learned almost immediately to approach the feeder whenever they heard it operate. As the meat fell from the feeder into their cage, they often reared up to swat it down in midair. If it hit the floor before they seized it, they would locate it by touch and smell, not by sight. Since both approached the feeder at the same time, there were occasional squabbles over the rewards. But within two hours, the latency of approach for both cats was less than one second, on the average. The next day, twelve hours after their normal ration of meat, average latency of approach for the small pieces of dried meat was still less than one second.

A metal paddle was installed on the ceiling of their cage. When the paddle was struck, it moved along a track. The harder a cat hit the paddle, the further it traveled and the more quickly it delivered a reinforcer (a thumb-sized piece of horsemeat).

I thought there would be very little stimulus generalization between the paddle and the things at which jaguars normally rear up to strike. To compensate for that lack, however, there was to be reinforcement for every instance of the behavior. Would the presence of reward compensate for a stimulus possessing very weak eliciting properties? *That*, to dramatize it just a little, was the question.

Initially, the track was mounted low so the cats would not have to rear to strike the paddle. The paddle was connected and the cats quickly habituated to it. The fact that it moved when touched did not interest them as much as I thought it would. Habituation was complete within thirty minutes, and my confidence in the compensatory ability of reward took a nosedive. After all, "moving when struck" was the *only* property which the paddle had in common with natural objects that might elicit striking and pursuit.

When the paddle was first installed, the cats made only cursory exploratory responses to it. A few sniffs within the first few minutes were their only responses during a two-hour period of observation. The prediction that the paddle would have very little control over natural rearing and striking movements was partly confirmed.

In view of their very fast habituation to the paddle, it was no surprise that shaping began very slowly. After an hour with the first cat, during which she did not even approach the paddle, I smeared some blood plasma on it. She immediately walked over and licked the paddle. As she licked, the paddle moved. It had gone no more than an inch when the reinforcer dropped into the cage. That first pairing of "move the paddle" and reinforcement was the only one that seemed in doubt. After that, it was just a matter of routine shaping. The same incentive worked for the other cat, and within a week the paddle was five feet off the floor and the cats had to rear up to move it.

At this point, the most intriguing feature of the jaguar project appeared. Both sisters alternated between pulling the paddle toward themselves and pushing it away. Pulling probably has a stronger innate probability of occurrence for a cat in most situations than pushing. But in this situation, pushing led to reward more quickly and slowly gained an edge over pulling.

But when the paddle was pushed into a corner of the cage, pushing postponed reward and led to momentary extinction. As pushing extinguished, pulling reasserted itself. Then, as soon as the paddle was pulled from a corner, pushing would reemerge as the dominant response. What threatened to become an impasse resolved itself without the arrangement of any special contingencies on my part.

The exhibit was beginning to serve the purposes set for it. The cats were getting about ten minutes of exercise per hour, four times a day. Frequency of aimless pacing was down and the public was spending a lot more time in front of the jaguar cage than formerly. The furtive glances cast my way during shaping were now replaced with overt expressions of interest. Exclamations of "Hey, kids, come over here and look at these tigers" began to increase in frequency.

I am always pleased when a trained behavior can occur without my intervention, because that is the point at which the animal becomes the focus of attention. That is also the point at which the institution can begin to tell an educational story about what is going on. Contrived though their behavior was, the jaguars were at least working for a biologically important consequence—food. And that is a step up from charity feeding with its frequent corollary—aimless, undirected activity.

All that remained was to give the animals a discriminative stimulus which told them when playing with the paddle would deliver meat. They could play with the thing anytime, of course, but to insure that the public would see them perform, it was necessary to teach the animals to respond on cue.

Fortunately, the cats extinguished very fast when reinforcement was withheld. And they reconditioned just as fast when reinforcement was present. Within three twenty-minute sessions, responding to the discriminative stimulus (blinking light) accounted for 85 percent of all responses to the paddle.

In retrospect, I think two other factors were operating to produce this rapid acquisition of light-on, light-off discrimination. The operant level of the cats' responses to the paddle was nil, indicating that it had practically no reinforcing properties of its own. Secondly, the response required a lot of effort. At the end of ten minutes, both cats would lie down next to the water dish, panting and drinking.

In order to assess the effects of training, I took five-minute time samples once an hour for five consecutive days. Pooling the data for both cats, I found that pacing was down to one hour (out of nine that the zoo is open) from a preproject figure of two hours; sleeping was down from two hours to forty-five minutes; eating was up to about one hour from a previous figure of fifteen minutes; time spent earning food was up to about an hour from a preproject figure of zero; and time spent reclining in an alert posture was up to ninety minutes from a previous total of thirty minutes. Other behaviors, like playing with the water hose, interacting with keepers, grooming, and interacting with each other did not change.

SURE, OUR OTTER SLIDES, UH, THAT IS, ONCE IN A WHILE

The next summer was the beginning of a 4-month involvement with otters. I had seen many otters in zoos, but had never chanced to work with one. I began reading Harris (1968), but could discover no obvious reason why so many otters in zoos did not use their slides. A half-dozen zoomen gave me conflicting explanations, agreeing only that increasing the frequency of sliding in zoos was an impractical goal.

Authorities on otters in the wild list sliding as a major part of the daily activity cycle. Emil Liers (1961), the otter man from Winona, Minnesota, wrote, "In the summer, otter will substitute mud or grassy banks for the snowy hills. But snow is always tops for otter slides [he is referring to the local species, *Lutra canadensis*]. . . . When traveling overland and on snow or ice they combine running with sliding. They run a ways, then fold their front feet under their bodies and slide; then they repeat the running and sliding process. Finding a hill that is clear of debris, they will coast to the bottom, do an Immelmann turn there, and run immediately to the top to repeat the performance."

Besides being a form of locomotion and a social-play activity, sliding may have some value in finding aquatic prey. Any behavior that is a basic form of locomotion, contributes to social cohesion, and may play a role in food-getting should be hard to eradicate from the repertoire. Nevertheless, I was told by one curator after another, "Sure, our otter slides. Uh, that is, I saw it slide once last year."

Long before I figured out that the answer lay in the field of ethology, I went to visit Emil Liers at the Otter Sanctuary in Winona. He talked about otters all one afternoon but did not say much about sliding. I was not surprised to see a child's playground slide leading into the deep section of the pool. "Sometimes they slide," said Liers. I asked him if I could try to train them to slide on cue. He agreed.

Using a part of their daily ration (cut pieces of calf's liver) as reinforcement and a metal dime-store cricket as a discriminative stimulus, I started learning about otters. Before I could do anything, both otters rushed up and began sniffing me over. This first exploration session lasted five minutes and was repeated three more times in the next hour.

The flowing stream of behavior issuing from those animals was an eyeopener for me. Seeing the vitality of Liers' otters, I puzzled more than ever about the lackluster otter exhibits at most zoos. The problem went beyond whether the animals slid or not. Liers' otters did not slide much either, but they did lots of other things.

For the next two hours, I tried to combine shaping and discrimination learning in order to get "sliding on cue." After twenty rewards, one of the animals would approach the steps to the slide, wait for the clicker, then slide. He slid in a sitting position, with forefeet braced in front and its rump down on the slide.

Lack of sliding in zoo otters is not due to captivity per se, nor necessarily to dietary deficiency, nor to poor physical condition. If any of those conditions were the reason, Liers' otters would not have slid so readily in so short a training session. In a general way, the reasons had to stem from the stimuli just preceding and just following a slide.

What might such stimuli be in the wild? The most important antecedents of sliding in the wild, aside from developmental age of the animal, are probably the physical features of the sliding surface. These features are not specified in anything I have read about otters, but one can reasonably presume that natural slides are concave, probably not over 40° in vertical angle, and are not strewn with large obstacles. Another important feature is that most natural slides extend right into the water. The slide at Liers' place was flat, tilted at about 50°, and ended two feet above the water. I could not recall one zoo among the fifty or so I had visited where the slide conformed to natural sliding surfaces.

As for the consequences of sliding, they include at least these: the social excitement that is generated within the group; moving from place to place (in winter); the experience of down hill movement; massage of the belly; the sudden contact with water; and perhaps an occasional meal of minnow or crayfish frightened into flight by the otter's entry into the water. Whether food reward is a major factor in natural sliding I cannot say, but I knew from working with Liers' otters that such reward definitely could become a controlling variable.

Not long after working with Liers' otters, I was given an adult female, small-clawed otter (*Amblyonx cineria cineria*). I put her in an outdoor pen twelve feet square with cyclone fence for sides and a floor of rock topped with sand. An oak tree for shade, a plywood sleeping box, and a 200-gallon horse trough completed her habitat.

She spent the first twenty-four hours sleeping outside her box on the sand. Early in the morning of the next day, I gave her a pan of lakewater containing some minnows and crayfish. She rushed to the pan and started drinking. The first time a minnow brushed her vibrissae, she jerked back violently, then plunged into the water headfirst. She ate all the minnows in a frenzy of movement, but ignored the crayfish. A few hours later, she ate her usual ration of ZuPreem.[1]

Only now, after sleeping and eating, did active adaptation begin. She sniffed the whole area, returning frequently to her sleeping spot for naps. Eating, sleeping, and exploration occupied her for four more days.

On the evening of the sixth day, she provided charming evidence that adaptation was nearing an end. As I came sliding down the rockpile toward her pen in the hollow, I heard a chirruping noise that sounded like a rowdy gang of oversized chipmunks. Coming closer, I saw Zeenie lunging at her tail and emitting this noise. She laid on her back, then assumed a dozen postures as the force of her lunging pulled her this way and that. She was oblivious to my presence, which I took to be another sign that she felt secure in her pen. From then on, she chased her tail every day, swam in the horse trough, and ran along the inside of her pen whenever I ran along the outside.

I wanted her to learn to go down a homemade slide. My main purpose was to demonstrate to zoomen that a captive otter could learn to slide on cue. As a study of behavior in a captive wild animal, however, certain other questions were also at stake. I have already mentioned the question of whether food reward would convert sliding from a genetically based behavior to a discriminated operant. If she did learn to slide on cue, the effects of food reinforcement on that behavior could be evaluated simply by comparing the number of reinforced slides to the number of unreinforced ones. Additional questions were whether she could find minnows in a murky body of water, whether she would satiate on live minnows, and whether the reinforcement of sliding would have any ill effects on the frequency of other behaviors.

---

1. ZuPreem is manufactured by Riviana Foods, Inc., Topeka, Kansas.

Magazine training began early next morning. In order to show her where the minnows would be coming from, I held a couple in a net on the chute that led from feeder to horse trough. She saw them jumping in the net, swam to the chute, and tried to climb it. As I released the fish, one of them fell against her vibrissae, causing her to jump backward into the trough. As she did so, the other fish slid into the tank. But she had sensed their passing because she dove immediately and came up with one in her jaws.

She dove, swam, and investigated the food chute for the next fifteen minutes. When she finally moved away, I released the minnows as before and, simultaneously, operated the feeder. As the feeder dumped one of its quart-sized buckets of water in which the minnows were held, she snatched a minnow from the stream that washed over her head, then dove for the fish that had slipped past.

She had eighteen more trials of magazine training in that first session. In the last twelve of these, her average latency of approach to the feeder was thirteen seconds. Average latency for the last six trials was less than ten seconds.

The slide, installed early the next morning, consisted of a wooden frame with a piece of plastic sheeting glued to it. The sliding surface was about eighteen inches across and slightly concave. The bottom rested on the edge of the horse trough about two inches above the water. To get to the top, she had to climb a steep ramp with wooden cleats nailed across it for traction. The vertical angle was about 25°.

She immediately rushed up to sniff over the slide. She galloped to the top of the ramp, with which she was already familiar, and put one tentative foot on the slide. Then she bolted down the ramp and into the water. It seemed as if the slide had frightened her, and the longer she stayed under water, the more I thought that. But then she surfaced at the bottom of the slide, pulled herself onto it, and began slowly climbing it, sniffing as she went. Halfway up, gravity took over. Her posture and facial expression as she slid slowly backwards into the water were the quintessence of what, in ourselves, we call amazement.

After that, I positioned myself about thirty feet from the pen and waited for a chance to reinforce some approximation of sliding. But she ignored the slide all the rest of that day. Next

morning there were muddy tracks on the slide. The day after that, I
saw her *walk* down the slide. Finally, on the fourth day after
installation of the slide, she ambled to the top of the ramp, peered
down, put her front feet gingerly on the slide, and skidded down on
four braced feet. Just before she entered the water, the feeder made
its hissing noise. She surfaced beneath it, caught one minnow as it
came down the chute, and dove for the other five.

But her sliding posture had been pretty exotic. She had gone
Liers' otter one better. That animal had at least sat its rump on the
slide. I checked Liers' description of sliding postures (1961, p. 32),
which read, "On a particularly steep and slippery slide the rare
otter that may not like the excessive speed will thrust his forepaws
forward to slow the pace or even brake itself to a stop. But it is
usually not long until that timid one has become accustomed to the
speed and he joins with the others in the sliding game with reckless
abandon."

There was hope. As long as she went down the slide, I was not
complaining about the topography of the response. As soon as she
caught the minnows from the first run, she jumped out of the water
and raced for the ramp. On this second slide, I began fading in the
auditory cue that was eventually to be the discriminative stimulus
for sliding. I waited until she had stepped off the ramp and onto
the slide before turning on the auditory cue. A second later, when
she was halfway down the slide, I turned off the cue and operated
the feeder. Its hissing noise (it was operated with compressed air)
was the last thing she heard before her head slipped under water.
She surfaced at the food chute and caught one of the minnows.

After her second slide, we settled into a routine. She slid for
minnows in two twenty-minute sessions a day. By the end of the
sixth session, I was presenting the discriminative stimulus some-
time *after* her arrival at the ramp, but *before* she started climbing it.
By the sixth session, the discriminative stimulus preceded her slides
by 3 to 5 seconds. Up to this point, she had made 9 unreinforced
slides (i.e., without the discriminative stimulus being presented)
and 48 reinforced ones. During the next week, she worked in eight
more sessions, making 19 unreinforced slides and 116 reinforced
ones.

The time between trials was determined by a variable interval
schedule ranging from 10 to 100 seconds, with a mean of 40

seconds. The discriminative stimulus had been moved back in time so that it was always given when she was at least six feet from the ramp and never given if she was already proceeding to the ramp. Her average latency to the discriminative stimulus during those eight sessions was 8 seconds, with time ranging from 3 to 25 seconds. Once the discriminative stimulus was turned on, she had up to 15 seconds to start her approach to the ramp. Otherwise, it was turned off. But the limited hold was unnecessary in Zeenie's case. No matter where she was or what she was doing[2], she bolted for the ramp whenever those pebbles rattled in the bread pan.

I continued training for another week, but the only changes consisted of a slight drop in the percentage of unreinforced slides (from 16 percent the first week to 12 percent the second) and a reduction in her average latency of approach to the cue (from 8 seconds the first week to 2 seconds the second week).

She swam in the tank at intervals throughout the day and as late as midnight, but I never saw her slide when not in a training session. Possibly she did, but such sliding could not have been frequent without my observation. Thus reinforcement became the only consequence that would support sliding in this otter in this situation.

One cannot avoid looking at a sliding otter. Even my farm neighbors, who are very busy in late August, began slowing down as they drove past. If the otter was sliding, they would leave their tractors on the road and walk down to the pen to watch.

Occasionally someone would ask if otters were not "supposed" to slide on their bellies. I always fumbled with that question, thinking it inappropriate to deliver a lecture on the difference between "unnatural" and "untypical." In a practical sense, I had not a clue on how to get Zeenie to slide in the typical posture. Probably the answer lay in some feature of the slide. The fear which Liers mentioned as a cause of sliding with braced forefeet seemed unlikely, because she bolted for the slide without hesitation when she heard the cue. Or perhaps a mild fear only came into play as the animal placed her feet on the top of the slide.

Some fussing with the slide would have been required to

2. The exception was scent marking. The discriminative stimulus did not interrupt that behavior. I am grateful to Nicole DuPlaix-Hall for identification of that behavior. I thought the animal just liked to scratch her genitals.

answer the question. The general reaction of my visitors helped me to recall that my goals for the otter were not research goals, so I settled for skidding on all fours.

My visitors also gave me fresh eyes to see what Zeenie was doing besides sliding, for my attention had been focused on sliding to the exclusion of much else. But visitors seemed fascinated by everything that Zeenie did, so I began trying to see what they were seeing. It was so obvious that I would surely have missed it.

What they were seeing between slides was swimming, diving, eating, scent-marking, grooming, napping, and exploration. As I began to record these additional behaviors, I saw that they usually occurred in that order. I thought back to what Liers and Harris had written, and slowly it dawned that she was emitting a *sequence* of natural behaviors. Sliding into the tank for live minnows was the key to the sequence.

Feeding in the water is the antecedent to grooming and scent-marking. These two behaviors only occur on land and their probabilities of occurrence following a period of absence from the animal's home territory are very high. The fact that her sleeping box was only a few feet from the water did not alter the relationship among these behaviors. The burst of energy required for her to run up the ramp, go down the slide, and pursue a half-dozen minnows is the antecedent condition for napping. Exploration naturally precedes food-getting—and while it is nonfunctional in Zeenie's case (i.e., the minnows are available whether she explores or not), it does occur as part of the sequence.

If the cues for sliding were spaced far enough apart, say 30 minutes, this entire sequence of behavior might be repeated many times a day. Visitors could be notified when to expect the otter to be active.

All the goals that I was beginning to think important for the display of captive animals—natural behavior, natural habitat, entertainment, education—could be achieved by selecting for reinforcement a behavior that would initiate a chain of other behaviors.

As regards the scientific issues previously mentioned, Zeenie had provided classic examples of pilot data (or "tease" data, as I prefer to call it). If one wished to pursue a program of research with otter, these data provided some justification for the following hypotheses:

1. Reinforcement for sliding increases the frequency of any form of sliding.

2. No matter how an otter gets down a slide, the behavior can be brought under stimulus control.

3. The typical form of sliding has both a genetic and a learned component, but otters can learn to emit this behavior under a variety of stimulus and reward conditions unlike those obtaining in the natural environment.

4. The form of sliding varies with the angle, concavity, and surface texture of the sliding surface.

5. Wild-caught otters of Zeenie's species do not satiate on a given prey species if bouts of catching and eating are spaced an hour or more apart.

6. The vibrissae and possibly other sense organs can substitute for vision in the pursuit of aquatic prey by *Amblyonx cineria cineria*.

WOODCHUCK IN A MUSEUM

The Science Museum of Minnesota, located in St. Paul, had a woodchuck (*Marmota monax*) which spent most of each day in a glass-fronted box. Occasionally it was taken to another room, plopped into a phony tree stump, then fished out by a narrator as she told a group of schoolchildren about "our friend, the wood-chuck."

Some discussions with the staff led to the idea of a stage drama featuring the woodchuck. It would emerge from its burrow as the stage lights came up and begin the day's feeding. When a stuffed fox appeared, it would rush down its burrow, etc., etc.

Enthusiasm was the order of the day for awhile, and I never doubted that the project could be done. Probably it can be, but my attempts to do it over a 5-month period were an unrelieved disaster.

I forgot all about the need to know an animal's personal history before planning a behavioral sequence for it. This wood-chuck had been hauled forcibly by the scruff in and out of its cage and that phony stump for three years. It was now between four and five years old. Its only response to the approach of any human was to hunch down and rattle its teeth. And I expected this sour-old-lady woodchuck to do a whole series of things on an exposed stage in the presence of forty wiggling, whispering schoolchildren.

Nothing worked. On stage, she was always hiding in dark corners or trying to jump onto the floor to find even darker hiding places. She was only moderately interested in food rewards, owing to a long history of being overfed. Because of my own timidity in reducing her weight, she was never in a proper state of deprivation. Magazine training took weeks under these conditions, and the latency of approach to the food cup varied wildly from a few seconds to eternity.

A coincidental visit to the Como Zoo sent the woodchuck project off in a new direction. I saw a yearling woodchuck being quite active on a log laid at an angle in an otherwise barren cage. It occurred to me that an ethological approach to displaying a woodchuck might make more sense than a theatrical one.

From that point on, we concentrated on building a habitat that would elicit some natural behavior from a young, wild-caught animal. The habitat featured a sloping surface twelve feet long and four feet wide. Beneath the surface was a burrow, complete with primary and secondary entrances. Part of an oak tree was set into the surface and a plaster-of-Paris tree stump looked convincing and also served to cover the secondary entrance. Walls were four feet high with the front wall made of transparent Plexiglas. The public could also see into the burrow through clear Plexiglas.

The Como Zoo loaned me their yearling male as a test subject for this miniature environment. As soon as he put foot into that space, there was constant action. Naps and bouts of grooming occupied the high end of the the surface, while toilet functions occurred at the low end behind the oak tree (in lieu of a toilet chamber beneath the surface). Dried hay dropped into the habitat was soon crammed into the animal's mouth and hauled into the nest chamber. The woodchuck slept in the burrow, but took catnaps on the surface. He avidly searched the surface for the little pieces of food scattered around, then obliged his audience by sitting up to eat them. He climbed and debarked the oak tree, ran in and out of the secondary entrance, jumped up on the tree stump, and even leaped to the top of the side walls where he put on a display of tightrope-walking.

The school kids now crowded around the exhibit on their knees to watch the woodchuck go in and out of the burrow. Their close presence did not bother the animal, who behaved toward the Plexiglas limit to his habitat the way he might have behaved to the invisible limit of his feeding range in the wild.

And that was the woodchuck exhibit. There is no objective data to report. Subjectively, the work provided me with an important practical lesson in the value of ethology for display of a captive animal.

ONE RIVER DOLPHIN

The Shedd Aquarium in Chicago, Illinois, has had a single river dolphin (*Inia geoffrensis*) on display since 1965. He is the favorite of the staff and interacts with dozens of them every day. Even so, he had virtually no stimulation from *within* his tank.

I began my work with this animal (named Chico) by teaching him two acrobatic tricks—a somersault and a backward tail-walk. Hand signals were the discriminative stimuli for these two behaviors. This part of the work was very casual, and no data exist on the number of reinforcements required to teach Chico those tricks.

Even at this early stage, the dolphin project was different from the others. For one thing, dolphins of the marine species had been giving public performances for more than twenty years. For another, Chico's history since capture was well-documented. And there is a vast literature on dolphins, including some reports on Chico's species.

Vocalizations for *Inia geoffrensis* are also well-documented (Penner and Murchison 1970; Caldwell and Caldwell 1967, 1970). A survey of its behavioral repertoire in captivity has been published (Layne and Caldwell 1964) and there is even a report of field observations (Layne 1958).

In counterpoint to these very encouraging background facts, there is the unfortunate circumstance that the Shedd Aquarium was designed for the display of fish. Although Chico lived in one of the institution's largest tanks, that tank was still only sixteen by ten by eight feet deep. No room exists in such small space for rocks, fallen trees, water plants, prey species, oozy bottom, or other typical habitat features. In short, the ethological approach to displaying Chico was not practical. Nor was it practical to put

Chico through an athletic performance of the Marineland variety. That was no great loss, however, since acrobatics had been done and done and done.

Tasks that could be accomplished in such a small space included vigilance (or reaction-time) behavior, matching-to-sample, and sonar demonstrations. The project began with vigilance, which I described to the staff in these terms: "Chico will have X seconds in which to detect a light and turn it off. His response to the light will turn it off and, simultaneously, turn on another one in ,a different location. The display as a whole will show scanning, rapid swimming and turning, and eating."

Before beginning the project, it seemed important to test Chico's vision. It would not do to install underwater visual stimuli and then discover that the animal could not see them. I also wanted to know the extent to which he employed sonar as an aid in discriminating underwater objects.

The simplest technique I could think of is called the *simultaneous discrimination method*. In this technique, an animal is presented with two stimuli on each trial. One of them is arbitrarily designated correct, which means that the animal is reinforced for responding to that stimulus. The other member of the pair is designated incorrect, and responses to that stimulus are never reinforced.

The stimuli were equilateral triangles (four inches on a side) and circles (four-inch diameter) cut from one-eighth-inch Plexiglas. Correct choices were reinforced by a click from a dimestore cricket, followed by a fish. The click had previously been paired with fish at feeding time and had acquired the property of a reinforcer.

Position of the correct stimulus was changed from trial to trial on a quasi-random basis. Each pair of stimuli was presented twenty-five times unless the animal made ten consecutive correct choices within the series. Chico left no doubt about his choice by gripping one of the stimuli in his jaws.

In the first four tests, the incorrect stimulus was always a clear, transparent circle while the reinforced (i.e., correct) stimulus was one of four colored triangles. When the triangle was yellow or white, he met the criterion of ten consecutive correct on the first ten trials. He did significantly better than chance when it was blue (seventeen out of twenty-five correct). But when it was red, his choices were not better than chance.

In the next three series, both stimuli were triangles. In the yellow-red and red-black pairs, he performed significantly better than chance. In the blue-white pair, his distribution of choices was only marginally better than chance.

These data allow no precise interpretation, but the next two tests may add clarity. In these, he had fifty trials in which to reverse a previous discrimination. Thus, between the yellow triangle and the transparent circle, the circle was now the correct stimulus. But he went to the yellow triangle thirty-eight out of fifty times. In the other reversal, it was the black triangle against the red one, with black now incorrect. He distributed his choices about evenly (twenty-seven and twenty-three) between the two.

His strong preference for yellow as opposed to the transparent circle, and his subsequent failure to reverse that preference, suggest a previously learned or even, perhaps, an innate preference.

In the next series, the stimuli were both black triangles. One was pointed up, the other pointed down. In fifty trials, he chose both triangles equally often. I was attempting in this series to eliminate vision and force him to make his choices on the basis of sonar. But, of course, he could have *seen* the difference in orientation of the triangles. In fact, however, he failed to discriminate the triangles with use of *both* vision and sonar. In the final series, he had to distinguish between a transparent circle and a transparent triangle. This test comes a little closer to elimination of vision, because transparent Plexiglas is much harder to see in the water than is colored Plexiglas. In fifty trials, he performed exactly at the chance level.

Head-swinging accompanied most of his choices, indicating that he was using sonar. If so, the form and size of those Plexiglas stimuli were not sufficient bases for discrimination. Alternatively, it is possible that his sonar was not very effective.

But I will not insist on any interpretation of these data, because on many occasions he simply charged for one or the other of the two stimuli without swinging his head. In such cases, if his choice was incorrect he headed straight for the opposite corner of the tank to await the next trial. It seemed to me that sheer lack of interest accounted for this behavior.

When he did approach slowly, swinging his head before choosing, he often showed a clear sign of annoyance if his choice was incorrect. His annoyance was expressed by releasing the stimulus, then biting it again much harder. Then as if to underline his feeling, he would sometimes surface and make a loud clapping noise with his jaws. I noticed this jaw-clapping in other situations which I judged to be frustrating, as when his daily massage was interrupted or when the tank man was called away during a feeding.

I now knew that he had good vision, especially for white and yellow objects. Consequently, the lights he responds to in his daily performances are bright yellow.

Training Chico to interact with the lights involved a series of steps. Shaping the basic response was the first and most difficult one. A stainless steel disc, about the size of a silver dollar, is the part of the manipulandum that Chico must push in order to turn off a light. Chico was willing to touch the disc, mouth it, and palpate it with his snout hairs—but his operant rate of pushing it was zero. As habituation to the plunger continued, even those other responses to it began to decline in frequency. It required three sessions and seventy-one reinforcers (pieces of thawed blue runner, his favorite food fish) to shape a response strong enough that he would activate the microswitch mounted behind the plunger.

During the next two sessions, a discriminative stimulus (a twelve-volt lamp mounted two inches above the plunger) was faded in. In these two sessions, he responded seventy times when the light was on (i.e., seventy correct responses) and sixteen times when it was off.

On the fourth day of training, two more identical manipulanda were installed. He now had to turn off two lights in succession to be reinforced. His response did not immediately transfer to the other two manipulanda. He did not point his snout directly at them on his approach, with the result that he often made contact with the side of the disc which did not register as a response. Another four sessions of shaping were required before he was equally adept at all three response stations. During this

Figure 5.1 Having just turned off three lights in succession, Chico picks up a fish. He frequently rolls onto his side or back when retrieving food items.

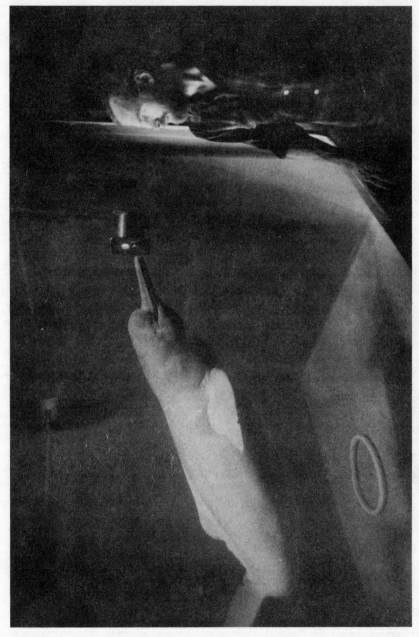

Figure 5.2 Chico in the act of turning off a light. The author watches from the public's side of the tank.

remedial work, he was reinforced for every correct response and deprived of the opportunity to respond (time-out) for every error. The total for each type of consequence for the four sessions was seventy-five reinforcers and twenty-five one-minute time-outs.

Chico now had no trouble turning off two lights in succession (a two-component homogeneous chain) before receiving a primary reinforcer. He did twenty-one of these chains in his first post-remedial session. On the afternoon of the same day, I required him to turn off all three lights to receive a reinforcer. He completed four of these three-component homogeneous chains immediately.

Next day, in the morning session, he did six three-component chains, and in the afternoon he completed twenty-five. I felt that chaining was mostly completed, so I began paying attention to his latency of approach to the first member of a chain. In that afternoon session just mentioned, the average latency was 6.8 seconds with a range from 3 to 10. That seemed a little long for a public display, so I instituted a limited-hold procedure.

In the morning session of the next day, Chico found that unless he turned off a light within nine seconds, it went out by itself. What's more, no other light came on for twenty seconds. This combination of limited hold and time-out was in force over the next eight days. As Chico's approach to a light became faster, the limited hold was gradually reduced to five seconds. The time-out remained constant at twenty seconds. During these eight days he was also, of course, practicing the three-component chain. On the last two days of this eight-day period, his average latency to the appearance of any light was four seconds. The range was from two to eight, but variability was narrowly restricted around four seconds.

And that was that. Thirty training sessions spread over eighteen days had brought the animal from inability to make the correct response to the reliable emission of a three-component homogeneous chain. That success speaks more for his intelligence than any hymn of praise in general terms could ever say. Remember that he had only one bit of information to utilize during the whole training procedure. If a behavior produced a fish, it was worth repeating; if it did not, it was not.

Since the work with Chico was done as much for the public's benefit as for Chico's, a questionnaire study was done to assess visitor reactions. Questionnaires were given to thirty-five visitors who saw Chico do his routine with the lights, and to forty-one visitors who saw the exhibit as it was before training. The results were weakly in favor of the trained sequence of behavior. The number of people standing in front of Chico's tank and the length of time they stayed were noticeably on the increase.

With this part of the project completed, the emphasis shifted to sonar. There was good behavioral evidence (i.e., headswinging) that Chico had used sonar in the two-stimuli choice tests, but I was not sure he made much use of it in his daily existence. He had no need of it, really, in a space where everything that happened was completely predictable. The question, as I interpreted it, was whether he *would* use his sonar. If he was rewarded for doing so, I thought he might use it frequently.

I started with a Clevite CH17 hydrophone plugged into a Sony TC124 tape cassette. It seemed sensible to conduct tests at night in order to eliminate visual cues and in hope that the unaccustomed hour would make Chico more likely to employ sonar to find out what was happening. Two sessions came and went without an audible squeak, although he did swing his head a couple of times.

A good chance existed that Chico was employing sounds outside my audible range. To test that hypothesis, the hydrophone was plugged into a recording oscilloscope for the next series of tests. When the ambient noise level in the tank, caused mostly by pumps and filters, appeared quite constant, a fish was dropped into the tank, then yanked out before Chico could seize it. As Chico reached the spot where the fish had been dropped, a burst of sound appeared on the face of the oscilloscope. The sequence of dropping a fish, then pulling it back, worked several more times and produced four photographs of energy bursts in the 75 to 120 kilohertz range. Chasing him around the tank with a long-handled net produced another four photos of vocalizations in the same range. As a bonus, Chico also produced about five seconds of low-frequency, audible sound. On the tape cassette, it sounded like the sudden movement of a rusty gate.

There is no longer any question that his sonar is intact. The hypothesis that external reinforcement is necessary for him to use sonar routinely during daylight hours has not been tested.

BEYOND RENOVATION

In addition to its use in changing extant behaviors and habitats, behavioral knowledge also has a role to play in the planning of new facilities. At the Minnesota Zoological Gardens, for example, considerable attention has been given to the relationship between habitat and behavior. In order to illustrate how the two fit together, I will summarize my report on the grizzly-bear exhibit planned for that institution.

The grizzly bear, as most readers know, is adapted to a wide range of habitats. The one chosen for grizzlies at the Minnesota Zoological Gardens is Rocky Mountain highland. No attempt will be made to reproduce such a habitat in complete geological and botanical detail, since only certain elements of the habitat are important for displaying natural behavior in the grizzly. The ethological literature suggests that these elements are water, sloping terrain, trees (fallen and standing), and heavy clumps of shrubs or dense evergreen stands (Craighead and Craighead 1963, 1965, 1971, 1972; Hornocker 1962; Martinka 1971; Mundy and Elook 1973). Properly placed in a one-and-a-half acre space, these few elements of grizzly habitat can support an extensive amount of natural behavior.

Once the supporting elements of the habitat are identified, the next problem is to *create* contingencies between certain elements of the habitat and the behaviors which grizzlies are genetically prepared to emit. For example, grizzlies will move rocks and logs in search of food. But if no food is found under those objects, they will soon quit moving them. Grizzlies will stand up to paw at bees' nests and bird nests, but only if they are occasionally successful in finding food in those places. They will fish in pools and streams only if they catch an occasional fish or frog. These contingencies already exist in nature. A grizzly on the prowl for food is on a natural variable-interval schedule. If he pokes this and lifts that, he is eventually paid off with a food item.

In a zoo, these contingencies must be created. Items of food can be automatically delivered to recesses beneath some of the rocks in the grizzly exhibit. A beehive made of textured concrete can hide a honey pump inside. Pools planned for the exhibit can support live fish.

Plantings in the grizzly exhibit have several functions. One is to provide visual barriers between dominant and subordinate animals. Another is to provide shady areas for lying-up. Still another is to give the female a reasonably attractive place to den. A fourth function is to simulate the appearance of a mountain slope.

The two pools currently planned also have definite functions. Grizzlies bathe, submerge, drink, and play in water. They also lie-up near water, especially if a food source is nearby, as in this proposed exhibit.

By combining the water, plantings, and hidden automatic feeders, a sequence such as the following may be expected: an adult male wakes from a nap, stretches, and yawns; he rises from his spot near one of the pools, stands up briefly, and sniffs the air; he ambles to a large fallen log and begins pushing one end; the log slowly moves and, after a time, the bear reaches into a recess beneath the log and comes up with a piece of carrion; he eats this en route to a large standing oak near the back of the exhibit; there he stretches to his full height and apparently tries to knock a bees' nest out of the tree; after a variety of pushing, pulling, and swatting movements, he cranes his neck upward and laps up the honey that begins oozing from the bottom of the nest; he then moves to the very top of the habitat where he looks at the holding pens hidden from public view behind a berm; satisfied, he now lumbers down to the second of the two pools; at his approach, two subadult bears rise from their resting place and begin retreating toward a dense clump of evergreens; with no more than a glance at the two retreating youngsters, the adult comes to the edge of the pool and stares into the water; his head swings this way and that for a few seconds, then he plunges his head into the pool, only to emerge a second later with nothing but water dripping from his head; he lunges for the fish several more times, but with no luck; he walks up to the other pool, shakes himself thoroughly, and resumes his former resting position.

Such a sequence might occupy as much as ten minutes and be repeated, in an endless variety of patterns, four or five times a day.

Females and young bears would have their own sequences as well. The bears, but not the public, will hear and see attenuated discriminative stimuli which will initiate the sequence of behaviors just described. The bears, of course, will also check out the places where food is delivered when discriminative stimuli are not present—and so much the better for display.

This sort of planning addresses two types of problems.

The first type concerns behaviors that we do not want to display: carnivores sleeping the greater part of their lives away, great apes staring blankly into space or reingesting their own vomit, canids trotting nervously back and forth, walruses swimming stereotyped patterns around their pools, polar bears swaying back and forth in place—the list is distressingly long.

The second type of problem addressed by contingency planning concerns all the wondrous behaviors we hope to see on a visit to the zoo or aquarium: fishing in grizzlies, armswinging in gibbons, social interactions among gorillas and orangutans, sliding in otters, leaping in tigers, herd defense in musk oxen—name your own favorite sight.

Any of these behaviors are possible for an institution to display if it will commit some of its resources to (1) discovering key elements of the natural habitat (not just pretty scenes, but factors known to be essential for a given behavior), and (2) creating contingencies between those elements and the natural behaviors related to them in the wild state.

Then the graphics and films and lectures and guided tours which describe food habits, social behavior, reproduction, food chains, climatic range of the species, and evolutionary history are supplements to, rather than distractions from, the exhibit. Then education and entertainment are predictable by-products of the exhibit. Then the viewer can experience something very close to the natural order. Surely, *that* is the point of displaying captive animals.

6 Development and Measurement of Cognitive Capabilities in Captive Nonhuman Primates

Susan M. Essock and Duane M. Rumbaugh[*]

WHY STUDY NONHUMAN PRIMATES?

EVOLUTIONARY PERSPECTIVE

The technological environment which humankind has contrived to inhabit is a most *un*natural one. As human beings, we have surrounded ourselves with an environmental morass of such alarming proportions that we must now struggle to survive our own cleverness. How might we modify our environments to reduce the stresses which threaten to thwart "civilized man"? To answer this question, we must first separate those influences predicated upon our biological, neurological, or genetic structure from those imposed by our environment. We must be cognizant of the extent of each of these classes of influence if we are ever to achieve an understanding of their interaction—our behavior.

How, then, do we recognize which of our perceptions, our social organizations, our *behaviors* are heavily influenced by our membership in the primate order, hence inherent in us as a species;

---

*Susan M. Essock was in the Department of Psychology at Brown University, and is now at the Wisconsin Regional Primate Research Center of the University of Wisconsin, Madison, Wisconsin. Duane M. Rumbaugh is Chairman and Professor in the Department of Psychology at Georgia State University, Atlanta, Georgia, and is also associated with Yerkes Regional Primate Research Center of Emory University. This research was supported by NIH grant RR-00165.

and which aspects of our being are contributed by the influences of our artificial environments? Since, on an evolutionary scale, our technological advances are very recent, a study of "primitive man" would allow us to see ourselves as we would be without the pressures of our current surround. This task is the primary objective of anthropologists' field studies of isolated tribes. They too try to tease out and define aspects of our nature which exist apart from what has been infused by our industrialized society. The comparative psychologist has chosen an alternative route by studying extant animal forms with whom we share common ancestors. By this reasoning, if we can define that which is *Primate* or *Simiae* or *Hominoidea*, we have come a long way toward defining what is human. Thus by studying both humans and other extant animal forms—and, in particular, our nearest living relatives, the nonhuman primates—we enhance the probability of achieving an evolutionary perspective of the course of events which produced the human population.

ENVIRONMENTAL CONSTRAINTS

At best, captive primates are housed by zoos and research institutions in artificial environments which attempt to approximate natural ones. Physical space and climate restrictions preclude the possibility of creating truly natural environments. But primates can be profitably studied even when they are housed in less than optimal conditions. For example, a captive primate group will still exhibit some patterns of social organization specific to their species; and once some of these social patterns have been recognized, it is possible to examine the effects of various environmental manipulations upon the social organizations. Comparison of measures of aggession in various captive environments with field reports of aggression can give insight into the pressures created by confinement. Since most humans live in a highly constrained environment, any information gained from studies concerning the pressures that crowding places on primate populations may help us find ways of coping with our own overpopulation problems.

A variety of captive settings also provides the potential for studying the effects of various early environments on later social interactions and learning skills. Such studies can investigate the role of early experience in cognitive development and address questions of obvious human concern. For example, if cognitive deficits accrue as the result of early impoverished environments, is this damage reversible; and, if so, what are the optimal conditions for counteracting the effects of a handicapping early environment? The more intelligent primates are ideal for this type of research because their cognitive skills are potentially great enough that any retardation of their development may produce a clearly definable difference in their subsequent test performance (e.g., Davenport, Rogers, and Rumbaugh 1973). Certainly humans desire the potential to structure a human infant's early environment so as to maximize its potential for social and intellectual development; and we would also like to have the ability to identify and hopefully counteract any adverse effects of less than optimal early conditions. Since the potential damage to subjects precludes the use of human infants for many studies from which valuable information for improving human rearing conditions could be obtained, we are fortunate to have great apes and other primates as animal models. In comparison to humans, nonhuman primates may be regarded as providing an array of "normal" retardates, with an array of cognitive skills which reflect varying degrees of cognitive development uncompromised by the physical handicaps or restricted rearing conditions which so often accompany the retarded human. As we learn how to enhance nonhuman primate cognitive processes through radical and/or high-risk methodologies, we should be better prepared for enhancing similar processes for the human child. Certainly, we must continually be cautious about extrapolating data to form hypotheses about one species from observations of another; but the fruits of such cautious research also give further insight into what is unique to the human species and what is shared with other primates.

Learning Capabilities

Along with providing information about the external factors which influence the ability to learn, primates (particularly the great apes) can be studied to investigate the mechanisms of learning. Certainly the creativity and insight demonstrated by humans, with our language capability and obvious talent for abstraction, has not been manifest to such a degree in any other primate form; but if one is willing to consider behavioral advances such as increased cognitive skills as the product of continually evolving processes, the mechanisms by which anthropoids and humans are able to learn should resemble each other by virtue of their common ancestry. Yet easily definable *qualitative* differences in the learning processes of various primate forms certainly appear to exist. This accords with the notion of continuity of behavioral mechanisms if we accept a position such as Nissen's when he observes that "quantitative complication may become so great that it produces, in effect, qualitative differences with new emergent properties" (1951, p. 351). An obvious example of an apparent "emergent property" which distinguishes the human species is our language capability. But less readily definable emergent properties also exist which appear to distinguish humans *and* great apes from other primate forms. Rumbaugh (1971) has presented evidence for qualitative differences in learning processes among primates—viz., that great apes appear to possess an abstractive ability not found in the lesser apes or monkeys studied. This matter will be further discussed in this chapter.

In many instances, the field behavior of free-ranging chimpanzees observed by van Lawick-Goodall suggests that many behaviors of these apes are more humanlike than monkeylike. Prior to these observations, tool manufacture and cooperative hunting ventures were thought solely to be manifestations of human evolution. Yet van Lawick-Goodall observed chimpanzees using chewed-up leaves as sponges to soak water and to cleanse themselves (1970), or modifying grass stalks to probe for termites (1971); she also reports the grouping of chimpanzees for a cooperative attack (1968) and calls attention to the remarkably familiar (to human and chimpanzee) gestures of begging with an open palm, greeting embraces, fear grimaces, and many others (1971). As Washburn and McCown (1972) point out, these similarities in motion result from similarly adapted anatomies. For example, humans and apes each leap and land with the hind legs, in contrast to the quadrupedal monkey style. Since so many of the behavioral categories previously assigned exclusively to the human repertoire have now been observed in great apes, we may be called upon to reevaluate the assumed boundaries of the cognitive domains which support such complex behavior.

As proposed by Le Gros Clark (1959), primate genera can be classified so as to increasingly approximate *Homo sapiens*. Beginning with lemurs through such varied forms as bush babies, tarsiers, monkeys, and lesser apes to the great apes, we can look for behavioral trends which may manifest various levels of cognitive or protocognitive development, which can then be examined as increasingly accurate *approximations* of human cognition. Obviously, such a chain should not be assumed to represent the evolutionary stages from which we evolved but may be regarded as evolutionary alternatives in which cognitive development was differentially emphasized, hence differentially developed (see Figure 6.1). But one of the most distinctive traits of the primates is this evolutionary tendency to develop a brain which is large in comparison to total body weight and is particularly characterized by an often richly convoluted cerebral cortex. This evolutionary emphasis on elaboration of the cerebral cortex will be referred to throughout this chapter as we look for behavioral manifestations of this "cortical development." Thus, by assessing the learning skills of primates very dissimilar to us as well as those of more similar forms, we can achieve a perspective concerning the refinements and embellishments which must have occurred to establish our current cognitive capabilities. Ideally, such a broad base of study will let us identify, assess, and deal with our capabilities and limitations in a much more accurate way than would be possible through the study of humans alone.

Washburn (1973) has summarized what he views as the "promise of primatology" into three major catagories:

(1) A fundamental challenge to the notion that there are different levels of knowledge and that the social should be studied without reference to the biological.

(2) A repudiation of the "black box" philosophy, and an insistence that the study of behavior must include an effort to understand internal mechanisms, including emotions.

(3) A repudiation of the thought that man is a rational animal and that the brain offers any simple road to truth. We too

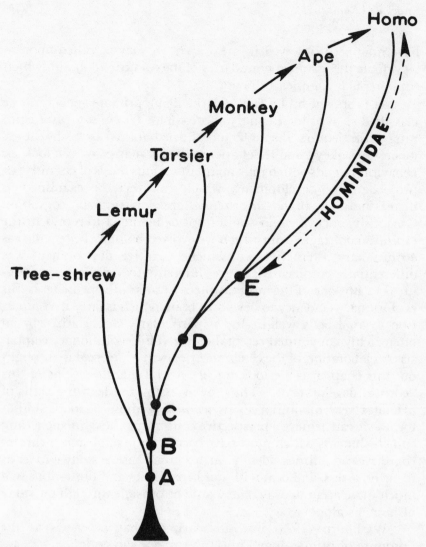

Figure 6.1 Diagram showing that living primates form a series from tree shrew to *Homo*, a series which suggests a general *trend* of evolutionary development. From comparative studies of such types, it is possible to postulate the probable *linear sequence* of hominid ancestry. This sequence leads from the basal primate stock through hypothetical transitional stages A-E, at which different ramifications of the primates are presumed to have branched off. Fossil primates approximating postulated ancestral stages are represented by crosses (from Le Gros Clark 1959).

are primates, carrying the advantages and limitations of our evolutionary history and indoctrinated in a particular primitive culture.

Washburn concludes by noting cautions supremely relevant to this chapter:

> Truth is a very restless thing. The promise of primatology lies in being animal oriented, problem oriented, and experimental. The less we trust the past, the more likely we are to be useful in the present. The more we can borrow techniques which probe far beneath superficial description, the more likely we are to make substantial progress. If we would understand the primates, we cannot accept the view of rational, scientific man which is deeply embedded in our culture. We are primates, products of the evolutionary process, and the promise of primatology is a better understanding of the peculiar creature we call man.

PRIMATES AS ENDANGERED SPECIES

Thus far we have endeavored to convince the reader of the value of primates, particularly the great apes, as research subjects. A word of caution should now be added for moral reasons. Many primates are members of endangered species. Although most apes should be able to survive and reproduce in captivity or in quasi-natural reserves, it is unlikely that poachers and human encroachment will allow them to survive much longer in their natural conditions. The mountain gorilla is one of many painful examples. Many mountain gorillas have been pushed higher up their slopes because of expanding human settlements. With a shrinking amount of forageable space, gorilla groups' ranges, which once rarely meshed, now frequently overlap (Fossey 1972). And, with vegetation on the higher slopes less suited to the herbivorous gorilla diet, their survival in the wild becomes still more precarious.

If future primate generations are the product of a very limited gene pool, we must assume that future generations may be morphologically and behaviorally different than those we know today. Care should be taken to protect existing species from extinction as we exploit them for all species' information we can obtain—for these may be the last examples of research subjects which, genotypically, are relatively uninfluenced by human intervention.

Methods for Assessing the Intelligence
of Nonhuman Primates

Intelligence as Learning Ability

Defined objectively, "learning ability" means the degree to which an organism can modify its behavior as a consequence of information gained through interaction with its environment. This, in less objective but more common usage, is also what we mean by "intelligence." When dealing with humans, we often appeal to the concept of intelligence to explain why some people achieve higher levels of education than others, get better grades in school, or excel on standardized tests. Similarly, if we are told that an individual is intellectually retarded, we expect a lower than normal performance level from that person on the tasks mentioned above. Formally, an intelligence measure is generally the product of a battery of tests, results of which have been found to correlate highly with one another and with other measures assumed to reflect intelligence. Individual tests are tasks of varying difficulty designed to challenge an individual's cognitive faculties to varying degrees. Examples of such tasks include recall of a string of numbers, vocabulary tests, or picture-completion tasks.

As the vagueness of these phrases might suggest, even those most skilled in psychometrics are generally loath to specify precisely what they believe intelligence tests measure; but there is agreement that all intelligence measures should provide a high correlation between a given score and the tested individual's ability to learn. Ideally, manual dexterity, eagerness for approval, or familiarity with similar tests should not affect the individual's score, which should primarily reflect the individual's *learning* ability.

In order to broach the question of estimating nonhuman intelligence, we will still require tests which measure a subject's ability to learn. As with measures of human intelligence, we want tests which remain unbiased by variations in eagerness of distractibility, or by the purely physical manipulative skills of the subject. A subject's performance level on such tests would be taken as a best estimate of its learning ability as compared to the other subjects tested. And hopefully, the relative performance scores would not vary as a function of the test employed. Two tests can hardly be measuring the same things if the scores on one fail to predict the approximate rank ordering of scores on the other.

LEARNING SET AND REVERSAL TASKS

We generally speak of learning as entailing a synthesis of information gained from one or more tasks which is then used on some subsequent task. Thus a change in a subject's performance which reflects use of additional information also reflects learning. The information assimilated may have been newly introduced or may have been present from the beginning of training. Harlow (1949) introduced a task, suited for obtaining performance measures, which could reflect the amount a subject was learning via the testing procedure being used. Harlow's task was a two-choice discrimination problem, where selection of one item in a given pair resulted in food reward and selection of the other item went unrewarded. Thus the first trial of such a discrimination provided information as to which object selection would result in food reward on subsequent trials. One pair of objects would be presented for a given number of trials followed by a pair of new objects for the same length-block of trials. Considering a subject's percentage of correct responses over many such trial blocks reveals how rapidly and how well the subject is learning to use the available information—how well the subject is "learning to learn" (Harlow 1949).

Harlow (1949) proposed that the most adaptive sort of learning, at least for primates, is "learning how to learn efficiently," and that "this learning to learn transforms the organism from a creature that adapts to a changing environment by trial and error to one that adapts by seeming hypothesis and insight." Since changes in a subject's performance parallel its formation of such *learning sets*, researchers are provided with an experimental tool with which variations in learning ability and faciltity can be observed and quantified.

Learning-set curves of trial-2 performance for rhesus monkeys (*Macaca mulatta*), squirrel monkeys (*Saimiri sciureus*), and marmosets (*Callithrix jacchus*) are shown in Figure 6.2. Initially, all subjects performed at chance; even after the first (information) trial, they picked the correct object only about 50 percent of the time. But effectiveness of the first training trial increased dramati-

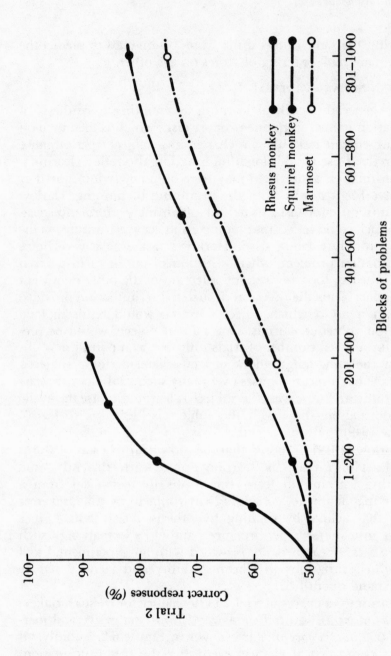

Figure 6.2. Improvement in trial 2 performance as a function of practice for the rhesus, squirrel monkey, and marmoset (Miles 1957).

cally with experience, as shown by progressive improvement in trial-2 performance following additional practice. The relatively steep rise in the rhesus monkey curve indicates that these subjects were rapidly learning to take advantage of the information provided by the first trial of each block. By the end of two-hundred blocks of ten trials each, the rhesus subjects had "learned to learn" which object was correct—from information provided on the first trial—to a degree sufficient to collect about 88 percent of the possible rewards from trial 2. The squirrel monkeys and marmosets apparently also "learned to learn," but to a lower asymptotic level than did the rhesus monkeys (indicating that the task was not learned as well by these two groups); and the squirrel monkeys and marmosets took longer to reach asymptote (indicating that these two groups learned what they learned more slowly than did the rhesus monkeys). Such cross-species comparisons, however, should be made cautiously and with appropriate reservations when no allowances are made for extra-learning-ability species differences. Such differences may confound learning ability with other species-specific variables that can affect test performance. Morphological differences, relative emotionality or motivational strengths (such as for food, social interaction, or sleep), food preferences, and the like may each serve to bias a certain procedure toward some species and away from others. Yet it is rarely possible for many different species to be tested in the same laboratory under the fairest possible conditions.

Comparative data gathered from various laboratories should be utilized, but cautiously; and Figure 6.3, showing comparative learning sets for several species gathered at various locations is here presented with reservations as noted. Notice the suggested, though perhaps equivocal, relationship between cortical development and learning-set acquisition as given in Figures 6.2 and 6.3. The Miles' results are as would be expected, considering that learning ability or intelligence is generally assumed to increase with cortical development (Noback and Moskowitz 1963; Conolly 1950), and that the rhesus monkey has a greater proportion of cortex to total brain than does a squirrel monkey, and similarly for the squirrel monkey to marmoset (Noback and Moskowitz 1963). A taxonomic tree such as that given in Table 6.1 also roughly serves to order the

Table 6.1 Summary of Primate Classification:
(Clover leaf from Schrier and Stollnitz 1971)
A Sample Taxonomic Tree

Suborder	Superfamily	Family	Subfamily	Genus	Vernacular name(s)
	Tupaioidea	Tupaiidae	Tupaiinae	*Tupaia*	Tree shrew
				Dendrogale	Smooth-tailed tree shrew
				Urogale	Philippine tree shrew
			Ptilocercinae	*Ptilocercus*	Pen-tailed tree shrew
	Lemuroidea	Lemuridae	Lemurinae (greater lemurs)	*Lemur*	Lemur
				Hapalemur	Gentle lemur
				Lepilemur	Sportive lemur
Prosimiae (prosimians)			Cheirogaleinae (lesser lemurs)	*Cheirogaleus*	Dwarf lemur
				Microcebus	Mouse lemur
		Indriidae	Indriinae	*Avahi*	Woolly lemur
				Propithecus	Sifaka
				Indri	Indri
		Daubentoniidae		*Daubentonia*	Aye-aye

Table 6.1 (Continued)

		Family	Subfamily	Genus	Common name
	Lorisoidea	Lorisidae		*Loris*	Slender loris
				Nycticebus	Slow loris
				Arctocebus	Angwantibo
				Perodicticus	Potto
		Galagidae		*Galago*	Galago (bush-baby)
	Tarsioidea	Tarsiidae		*Tarsius*	Tarsier
Simiae		Callithricidae		*Callithrix*	Marmoset
				Leontocebus	Tamarin, pinche
			Callimiconinae	*Callimico*	Goeldi's "marmoset"
	Ceboidea (New World monkeys, platyrrhine monkeys)	Cebidae		*Aotes*	Douroucouli (night monkey, owl monkey)
				Callicebus	Titi
			Cebinae	*Pithecia*	Saki
				Chiropotes	Saki
				Cacajao	Uakari
				Alouatta	Howler monkey
				Saimiri	Squirrel monkey
				Cebus	Capuchin
				Ateles	Spider monkey
				Lagothrix	Woolly monkey

Table 6.1 (Continued)

		Family	Subfamily	Genus	Common name
Simiae	Cercopithecoidea (Old World monkeys, catarrhine monkeys)	Cercopithecidae	Cercopithecinae	*Macaca*	Macaque
				Cynopithecus	Black ape
				Papio	Baboon, drill, mandrill
				Theropithecus	Gelada
				Cercocebus	Mangabey
				Cercopithecus	Guenon
				Erythrocebus	Patas monkey (hussar monkey, red monkey)
			Colobinae	*Presbytis*	Langur, leaf-monkey
				Pygathrix	Douc
				Rhinopithecus	Snub-nosed monkey
				Simias	Pig-tailed langur (Mentawi Islands langur)
				Nasalis	Proboscis monkey
				Colobus	Guereza
	Hominoidea (apes and man)	Hylobatidae (lesser apes)		*Hylobates*	Gibbon
				Symphalangus	Siamang
		Pongidae (great apes)	Ponginae	*Pongo*	Orangutan
				Pan	Chimpanzee
				Gorilla	Gorilla
		Hominidae		*Homo*	Man

174

Mean percent correct on trial 2

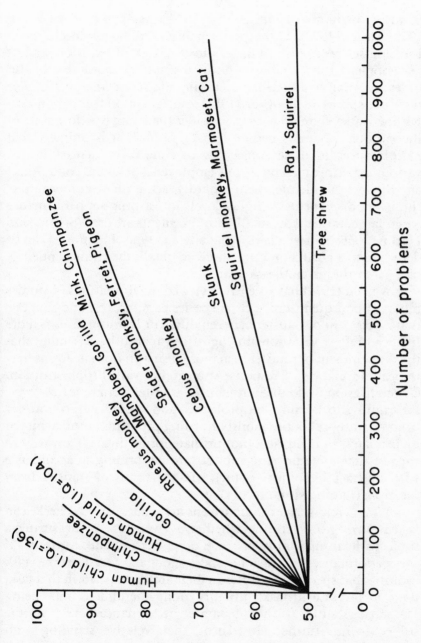

Figure 6.3 Improvement in trial 2 performance as a function of practice for various species (Hodos 1970).

175

primates in terms of increased cortical capacity (Le Gros Clark 1959; Mason 1968). If the performance of different species on a learning-set task tends to order those species in terms of cortical development, then, by the reasoning discussed, the task should also order the subjects in terms of learning ability or *intelligence*. For the three species used, Miles' (1957) results support this hypothesis, but the reader should beware. Consider the relative performance of the various species as depicted in Figure 6.3. It is unlikely that performances of these subjects were equally influenced by the various experimental conditions employed. "Standard conditions" are difficult to define, hence hardly achievable. For example, rhesus monkeys achieve higher levels in learning-set performance when large, rather than small, reinforcements are used (Schrier and Harlow 1956; Schrier 1958). What size and type of reinforcer can be delivered to a gorilla or a tree shrew to obtain the equivalent effect of a given rhesus monkey reward?

Warren (1974) has recently reviewed the limitations of studies relying upon quantitative cross-species comparisons and suggests that assessment of transfer between different learning tasks is more likely to reflect variations in cognitive capabilities. He notes that the development of transfer seen most clearly in primate forms may indicate formation of response strategies of varied sophistication. Generalization of double alternation, transfer from repeated reversal training to learning-set problems, and suppression of transfer from learning-set to probability learning problems are all achieved by primates, in contrast to nonprimate forms, in such a way as to suggest that some primate subjects are performing in accordance with abstract hypotheses which are the result of transfer from multiple training situations.

When cross-species comparisons are made, various checks can be employed to determine whether a given task is measuring a subject's learning ability or some other performance attribute. If the performances of various species on a given learning task indicate that species may differ in their ability to perform that task, then an increase in task difficulty along the dimension(s) being measured should further separate performance levels of the subjects—thus further elucidating their relative standing with respect to the particular abilities being measured. A reversal task is just such a refinement of the basic learning-set procedure. For example, Harlow (1949) ran subjects on a given discrimination problem for seven, nine, or eleven trials, then reversed the reward contingencies so that the previously positive stimulus became the unrewarded stimulus, and selection of the previously negative stimulus was reinforced. Thus, as with the learning-set task, an improvement on trial-2 performance after the reversal trial (which provides the information that reinforcement contingencies have been reversed) indicates that the subject has learned how to take advantage of new information. Harlow points out that by the end of reversal training, his subjects (rhesus monkeys) possessed learning sets for both learning and reversing a given response, and that the appropriate behaviors were emitted immediately and with "hypothesis-like efficiency."

The reversal task is more complex than the basic learning-set task because, in order to gain reward, the subject must not only take advantage of new information, it must specifically *not* respond to previously reinforced stimuli. Thus it was with some surprise that Rumbaugh and McCormack (1967) reported that macaques (*Macaca niger, M. silinus,* and *M. nemestrina*) generally did better than great apes (*Gorilla, Pan, Pongo*) on learning-set, reversal, and oddity problems. Fortunately, in reconsidering their data, Rumbaugh and McCormack noticed two trends not emphasized in the original study: that trials to learn an initial discrimination problem could *not* be used to predict reversal test performance on subsequent problems; and that relative performance on reversal trials *was* related to cortical development as long as prereversal performance accuracy was taken into account.

We mentioned earlier that a host of species-related variables could plague experimental research; sensory, motor, and attentional differences, emotionality, and varying motivational strengths (for food or competing motivations) all serve to bias a test situation based on a fixed number of trials in favor of some species and against others. In contrast, such biases are circumvented by testing only after a particular pretest performance criterion has been reached, thus insuring equivalent mastery for that problem which then *may* be transferred from problem to problem in a cumulative manner; i.e., with reference to Harlow's "learning-to-

learn" description of learning set, criterional training equates
subjects on what they have learned in order to achieve a comparison measure for how well they are "learning-to-learn" more.
Relative reversal performance following criterional training
should then only reflect differing abilities to learn the reversal task
attributable to differences in complex-learning capabilities.

TRANSFER INDEX

As an attempt to develop a species-fair test of learning ability,
Rumbaugh (1969, 1970) formulated the Transfer Index (TI). TI
scores are obtained by training subjects to a given criterion (either
67 percent or 84 percent correct) on each of a series of two-choice
discrimination problems similar to those used in standard
learning-set studies. When the criterion is reached, the cue values
for that problem are reversed for the next ten trials. The TI is the
ratio of percentage correct on reversal trials to percentage correct
during the criterion period of acquisition (TI = R%/A%). Within a
criterional level, A percent is fixed because the subjects are trained
to a specific criterion; thus the TI varies as a function of percentage
reversal correct. This means that subjects can vary widely in their
speed to reach criterion without affecting their TI score, and a
subject's TI score will not suffer because of the time possibly
required to adjust to the experimental situation or to learn the
initial task. Generally, the number of trials to reach criterion for a
given problem is not predictive of the reversal performance on that
problem. As seen in Table 6.2, correlations between trials to
criterion and reversal performance per problem are typically
slightly negative and rarely reach significance. For TI comparisons
to be meaningful, these correlations must be slight because, by
equating subjects on a performance criterion, the TI methodology
assumes that the number of trials to criterion will not influence
subsequent reversal performance. Both Rumbaugh and Gill (1973)
and Rumbaugh (1970) results support this assumption.

Rumbaugh and Gill (1973) used TI methodology to test forty-
five great apes (fifteen each of *Gorilla, Pan,* and *Pongo*), six lesser
apes (*Hylobates*), four vervet monkeys (*Cercopithecus*), and four
lemurs (*Lemur*). Two TI scores for each subject were first obtained
at the 67 percent criterion, then two additional TI scores were
obtained using the 84 percent criterion. Results of these tests are

Table 6.2 Pearson Product-Moment Correlations (r) Between Trials to Criterion and Reversals Performances (Per Problem) Obtained with TI Testing Procedures (Rumbaugh 1970)

		Pre-reversal 67%	Criterional Level 84%
Species			
	Orang-utan	.22	.30
	Gorilla	.34*	.39**
	Chimpanzee	.11	.17
	Gibbons	.30*	.53**
	Vervets	.08	.09
	Talapoins	.10	.10

*p .05
**p .01

shown in Figure 6.4. General superiority of the great apes over the lesser apes and monkeys is apparent. The great apes were the only groups to show positive transfer between the training and reversal trials (as evidenced by TI values over 1.15 for 65 percent criterional training). As the prereversal criterion was increased to 84 percent correct, the reversal performance of each of the great ape groups improved commensurably by about 18 percent, roughly the difference between the 67 percent and 84 percent performance levels. This is reflected by the relatively constant TI scores for the great apes at either criterion. The vervet group also displayed a positive relationship between criterional prereversal level employed and reversal performance; however, both vervet averages were lower than any of the great ape averages. Apparently subjects in each of these groups were able to form learning sets which facilitated their reversal performance, and the benefit of these sets to performance was greater when the prereversal task was learned best (at the 84 percent criterion rather than the 67 percent criterion).

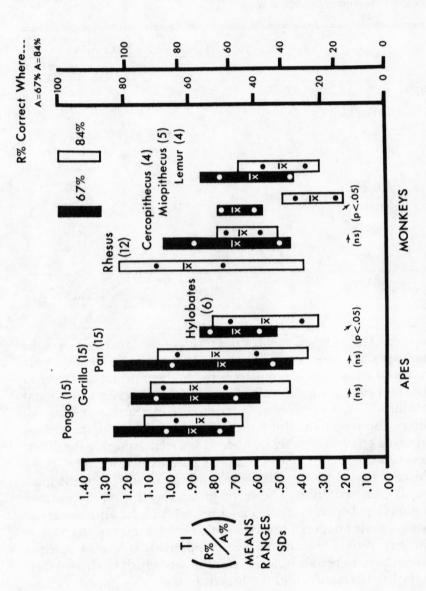

Figure 6.4 Transfer Index values for apes, monkeys, and lemurs (Rumbaugh and Gill 1973).

In contrast to a positive learning-set transfer as the result of a higher pretest performance criterion, it is possible to conjecture that improved learning of the prereversal task might make the reversal task more difficult to learn. The talapoin subjects behaved this way; the better they had learned which object was initially correct, the more they continued to choose the previously positive object after the reward contingencies were reversed. For the talapoins, there was a negative relationship between prereversal criterional level required and reversal performances. Gibbon and lemur reversal performances changed little with an increase in the prereversal training criterion, hence each group suffered a decrement for reversal performance relative to acquisition performance—this is reflected in the lower TI scores at the 84 percent training criterion relative to the 67 percent training criterion. Figure 6.5 shows the reversal percentage correct for each training criterion for each group. Each group evidenced at least some learning of the reversed problem, if not of the reversal task. If the subject did not attend to the reversal and continued to apportion its responses on the prereversal basis of 67 percent to one object and 33 percent to the other (or 84/16 percent, depending on the training criterion used), reversal performance should have been about 33 percent correct (or only 16 percent correct for the A percent = 84 percent series). All groups performed above this level, indicating that they responded to the shift in reward contingencies at some point following the first reversal trial. The sharp increase in percentage correct for the great apes immediately following the reversal-information trial and after 84 percent criterional training suggests that they did form sets to reverse (Figure 6.7). In contrast, the gibbons' gradual increase in reversal percentage correct suggests an incremental sort of learning in which there was gradual extinction to the previously positive cue and/or counterconditioning to the previously negative cue (Figures 6.6, 6.7). The great ape post reversal performance after 67 percent criterional training falls somewhere between these extremes (Figure 6.6); post reversal trial 1 is only about 37 percent correct—about what would be expected on the basis of 67 percent criterion training—but the reversal percentage-correct curve climbs rapidly, then levels off and does not exhibit the gradual rise seen in the gibbon curve. Again, the relatively rapid change in responding demonstrates the superior capabilities of the great apes for transfer of training skills to novel, discrete situations.

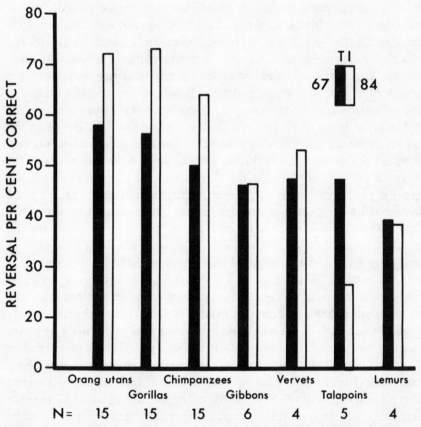

Figure 6.5 Reversal performances obtained from the Transfer Index testing of Figure 6.4 (Rumbaugh and Gill 1973).

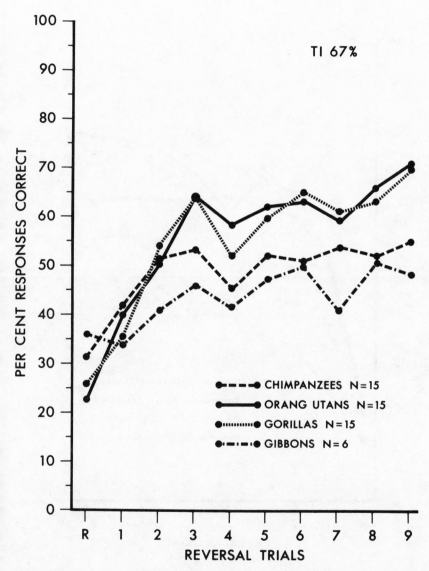

Figure 6.6 Percentage responses correct per reversal trial for apes as obtained from Transfer Index testing with the prereversal criterional mastery level of 67 percent correct responses (Rumbaugh and Gill 1973).

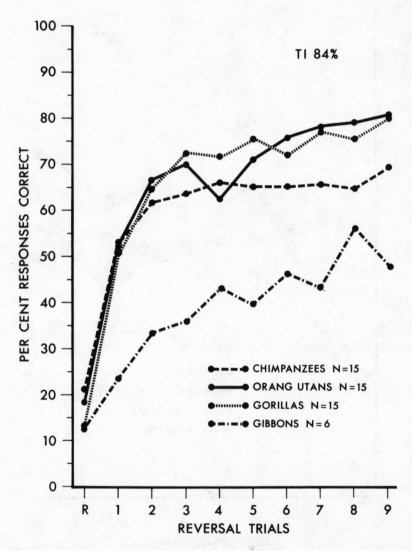

Figure 6.7 Percentage responses correct per reversal trial for apes as obtained from Transfer Index testing with the prereversal criterional mastery level of 84 percent correct responses (Rumbaugh and Gill 1973).

Learning Ability Differences Across and Within Species

The studies thus far discussed are typical of the primate-learning literature in their suggestion that cortical development serves as a rough predictor of the relative performance of various species on learning tasks. Notable exceptions to this trend are the poorer-than-expected performance of lesser apes (gibbons) and the frequently superior performance of various macaques, but such are exceptions to an otherwise general rule. In the Miles study (1957, Figure 2), no squirrel monkey performed as well as the poorest performing macaque. Performance differences across families are typically large, possibly indicating qualitative differences in learning abilities which are reflected by large performance differences. Generally, apes can be expected to perform better than monkeys, which in turn can be expected to do better than prosimians. When such an ordering is not achieved, experimenters should examine the task and search for variables which may favor one species or disadvantage another, because such susceptibilities often provide insights into species-specific behaviors which might otherwise pass unnoticed.

Performances within the great ape family are generally commensurate with one another. Rumbaugh and McCormack (1967), Rumbaugh (1970), and Rumbaugh and Gill (1973) found no consistent differences between the performances of chimpanzees, gorillas, and orangutans. Such equivalence of performance among the great apes is contrary to some unfortunate zoo lore. Chimpanzees are often reputed to be the "smartest" of the apes, and orangutans have the reputation of being dull and sluggish. Such tags are unfortunate and contrary to the results of studies presented here. Much of the lore probably stems from the chimpanzee's tendency to mimic (ape) human behaviors. We consider ourselves marvelously clever; ergo, so is the chimp. Gorillas poke things with their fingers that chimps slap with their palms. And orangutans will, likely as not, pick at the screw *next* to the reward. At the end of a test session, an orangutan or gorilla may be slowly dissolving a mouthful of M&M candies, while a chimpanzee will not. Such variations in temperament and distractibility can lead to erroneous conclusions about the relative *capabilities* of these animals, whether these conclusions are based on biased testing procedures or on unfortunate human conjectures.

The frequently superior performance of the macaques as compared to apes may be due to extra-learning-ability factors, since numerous other tests indicate that the ape is indeed the superior learner. (For a review of the learning skills of the rhesus macaque, see Rumbaugh and Gill 1975.) Instances of tool use and cooperative behavior (as well as the communication skills to be discussed) widely reported in apes, but minimally apparent or absent in monkeys, support this claim. Cross-modal perception is more readily demonstrated in apes than in monkeys. Davenport and Rogers (1970a) first demonstrated that orangutans and chimpanzees can discriminate between two objects solely on the basis of tactile cues, then select that object which matches a visually presented sample. This ability to abstract and exchange information between different sensory modalities was often assumed to be an exclusively human ability, possibly contingent upon availability of a symbolic language to mediate the cross-modal information (e.g., Ettlinger 1967). Clearly, language mediation is unnecessary to perform this task, since alinguistic apes can match haptically presented (i.e., touched) stimuli to a visual sample (Davenport and Rogers 1970a). Apes can also make successful visual-tactile cross-modal matches from photographs to real objects (Davenport and Rogers 1971); and bidirectionality of the visual-tactile cross-modal perceptions has also been demonstrated in apes (Davenport, Rogers, and Russell 1973). Apes have been trained to a criterion of eight correct out of ten trials after as little as 3 months of training. In contrast, rhesus-macaque subjects studied under these same conditions and trained by the same staff can match a haptically presented object to the correct visual stimulus with only slightly better than chance accuracy after more than three years of training.[1] This is not to say that rhesus monkey subjects will be unable to match cross-modally with a high degree of accuracy, but that their performance to date is far inferior to those apes studied.[2]

---

1. Davenport 1974: personal communication.
2. Since this chapter was written, Alan Cowey and Lawrence Weiskrantz (*Neuropsychologia*, 1975, *13*, 117–20) have demonstrated cross-modal perception by rhesus monkeys. The monkeys first learned on edible versus inedible shape discrimination in the dark (by touch). Subsequent tests in the light showed cross-medial transfer from touch to vision.

A series of studies by Gallup and his co-workers has also provided a performance difference between apes and monkeys (Gallup 1970; Gallup and McClure 1971; Gallup, McClure, and Hill 1971). In these studies, chimpanzees with a few days of prior exposure to their reflected images in mirrors learned to recognize their own images. Although they initially reacted to the mirror image as if the image were another animal, social behaviors such as bobbing, threatening, and vocalizing waned sharply over the first three days of exposure; and behaviors directed toward the self, apparently via the mirror, increased. The chimpanzees were observed to look into the mirror when grooming otherwise inaccessible body parts, while picking bits of matter from their teeth or nostrils, or while observing anal-genital areas. They also "made faces" into the mirror or sat and observed themselves blowing bubbles. When part of an eyebrow and an ear of mirror-sophisticated chimpanzees were marked with a tasteless, odorless red dye, instances of touching the marked body parts rose dramatically over pretest levels. In each of these instances, the self is the referent through the reflection, whereas in cases of social behavior directed toward the mirror, the reflection is the referent (Gallup 1970). Under similar conditions, rhesus and other monkeys as well as isolate-reared chimpanzees continued to react to the mirror image as if it were another animal. These studies appear to demonstrate a concept of self for normal chimpanzees and suggest that a concept of self may be attributed to humans and apes but may not extend to other primates.

The often small quantitative differences observed between monkeys and apes on more traditional behavioral tasks may also be due in part to extra-learning factors. Monkeys generally work for food much more readily than apes. A hungry ape can be a very poor subject, whereas increasing food deprivation will often improve a monkey's attentiveness and readiness to respond. Competing stimuli such as the sight of a caretaker, a bug, or the squeak of a door are more likely to divert the attention of an ape than a monkey subject. It is quite possible that the learning-set type of task is, in itself, boring to the ape, and that ape versus monkey performance differences will only be seen when tests employed are sufficiently difficult so that neither group will perform near one hundred percent correct.

Gibbons and orangutans appear particularly susceptible to potential distractions which do not disrupt the performance of other species operating under the same test conditions. Rumbaugh and McCormack (1967) noted that the learning-set formation of some subjects may have been compromised because the test objects were encased in Plexiglas bins. The more arboreal subjects were more easily distracted by irrelevant foreground cues (such as one-half-inch wire mesh inserted behind the Plexiglas but in front of the object) than were the less arboreal forms. A subsequent study using gorillas, chimpanzees, and orangutans has supported this hypothesis; introduction of such a screen disrupted orangutan performance significantly more than gorilla or chimpanzee performance (Rumbaugh, Gill, and Wright 1973). Attentional biases, such as this tendency to look only at what is in the visual foreground, can often be overcome by criterional training. High-level learning-set performance has been coaxed from a gibbon by first teaching the subject a single problem, then introducing various irrelevant cues, retraining to the criterion each time. When subsequently tested on 100 new problems, the gibbon performed with 91 percent accuracy, commensurate with the best of the great apes tested (Rumbaugh and McCormack 1967).

We have seen that various primate species perform learning-set and reversal tasks with widely differing accuracies. Do these performance differences reflect quantitative or qualitative differences in the learning processes of these subjects? This question must be broached carefully, for even widely disparate performances need not imply qualitatively different processes, and identical performances do not warrant the assumption that the subjects have learned via identical processes. If qualitative differences in learning processes do exist among nonhuman primates, we would expect the primates which differ most in cortical development to manifest these qualitative differences.

Rumbaugh (1971) presented evidence for qualitative differences in learning processes of gorillas, as compared to gibbons or talapoin monkeys, using a modified discrimination-reversal paradigm. Following criterional training on each object-quality discrimination problem, subjects were presented with one standard reversal trial (A+ became A– ; B– became B+) and then ten additional trials, during which one of three possible procedures was followed: reversal trials could continue in the usual manner (A- , B+), or a new stimulus could be substituted for either A or B (C- , B+ or A- , C+). Since these test conditions carried forth either A or B or both A and B from trial 1 of reversal, enough information was always present for a subject to perform at one hundred percent on post-reversal trials. If a subject was functioning under a hypothesis generated from abstract learning (for example, some refinement of Levine's win-stay, loose-shift strategy), these three conditions might be of equivalent difficulty. If, however, the subject's learning was via excitatory conditioning to the positive stimulus and/or inhibitory conditioning to the negative stimulus along the lines suggested by Hull (1943) or by Spence (1956), then performance in either condition involving the new stimulus C should allow for better reversal performance than the conventional reversal condition (A- , B+). After reversal in the A- , B+ condition, a subject that learned the original discrimination by habit accrual, or by some other means of increasing the associative strength of stimulus A, would first have to undergo extinction to the previously positive stimulus, then counterconditioning to the (possibly) inhibitory stimulus B. Since either extinction or counterconditioning is avoided when a new stimulus C is substituted either for A- or B+, either of the C conditions should allow for better reversal performance than the conventional reversal procedure of A- , B+.

Five each of gorillas, gibbons, and talapoin monkeys were tested using each of the three post-reversal procedures. The gibbon results were ambiguous because of the strong preference of these subjects for the novel stimulus C, regardless of test condition. But the differences between gorilla and talapoin results were striking because of the equivalence of gorilla performances on the three reversal conditions, whereas the talapoins found the A- , B+ condition significantly more difficult than either condition involving substitution of a novel stimulus for A- or B+ (Figure 6.8). These results support the assumption that qualitative differences do exist between the learning abilities of gorillas (great apes) and talapoin (Old World) monkeys. Talapoin performance is consistent with the assumption that they learn by S-R association, whereas the gorilla appears to use abstract rules such as: given a loss on reversal trial 1 because of a choice of A, either (1) if the previously reinforced stimulus is present (here A), shift to any other

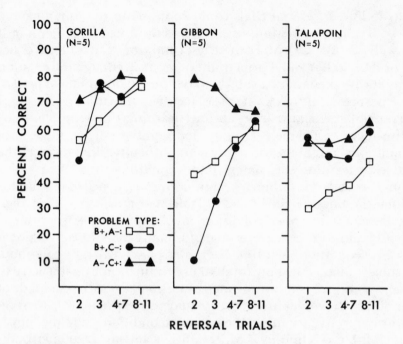

Figure 6.8 Reversal performances for three primate groups on each of three types of discrimination reversal problems (Rumbaugh 1971).

alternative (here B or C); *or* (2) if the previously unreinforced stimulus is present (here B), shift to it. Only the gorillas' performance is consistent with such a conditional discrimination. These results are in agreement with the views expressed years ago by Köhler (1925) and Yerkes and Yerkes (1929): that great apes are particularly adept at abstraction of information into general rules when compared to other nonhuman primates.

We have just reviewed evidence for qualitative differences between various primate species. Does evidence exist for qualitative differences in learning processes *within* a given primate species? Certainly, frequently large quantitative intraspecies differences in performance are observed (e.g., Figure 6.4). Are these performance differences produced because some subjects within a given species possess different learning skills—such as the ability to form abstract hypotheses—or because subjects differ in the extent to which their equivalent learning processes can be behaviorally manifested?

To address this question of possible qualitative learning-process differences within a species, Gill and Rumbaugh (1974) tested two groups of four adult apes (two chimpanzees, *Pan*, and two orangutans, *Pongo*, in each group), following the experimental procedure just outlined for investigating qualitative differences in learning processes among gorillas, gibbons, and talapoins (Rumbaugh 1971). The groups were labeled "bright" or "dull" on the basis of prior transfer index (TI) performance; average TI for the Bright Group was 1.00 and for the Dull Group 0.68, using a 67 percent correct criterion during acquisition. These values are very similar to those found by Smith (1973) when she tested two groups of four-and-a-half-year-old humans who had mental ages averaging about two-and-two-thirds years old for the low IQ group and five-and-three-fourths years old for the high IQ group.

Gill and Rumbaugh supposed that the Dull Group might lack abstractive powers present in the Bright Group which would make

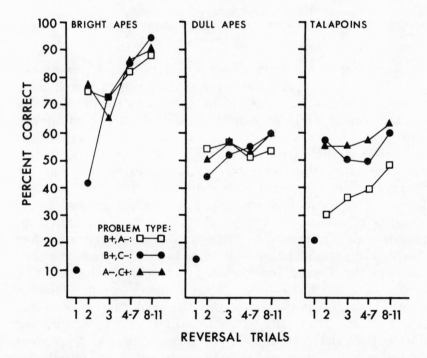

Figure 6.9 Reversal performances for three primate groups on each of three types of discrimination reversal problems (Gill and Rumbaugh 1974).

the reversal tasks involving the novel stimulus C simpler.
Although the Dull Group performed consistently lower than the
Bright Group, there was no difference between post reversal test
conditions (A- , B+; A- , C+; or C- , B+) for either the Bright or the
Dull Group (Figure 6.9); in keeping with the interpretation that
each group learned abstractly during criterional acquisition, there
was no species interaction. The groups did not differ in their
average number of trials to reach the prereversal criterion (10.8
trials for the Bright Group and 11.4 trials for the Dull Group), yet
the postreversal tests showed that the Dull Group was indeed
retarded when compared to the Bright Group. Gill and Rumbaugh
concluded that the phenomenon of mental retardation is not
unique to humans, and that investigations of the mechanisms and
causes of mental retardation may be carried out on nonhuman
primate subjects with possible applications toward combating
human retardation.

Environmental Influences on Cognitive Development

Deprivation Studies

Because the detrimental effects of restricted rearing conditions
upon human social and cognitive development are well-documented and of enormous humanistic concern, we might expect this
area to be one in which primate research would offer great value. It
is frequently impossible, for ethical reasons, to use human infants
as subjects in studies where the experiences of one experimental
group are restricted in comparison to another experimental group,
especially since we would expect the restricted group to suffer
developmentally. In such situations, an adequate animal model
would be invaluable. Unfortunately (but perhaps typically), the
primate species used as a model for human development must be
chosen with great caution, for research with various primate
species on the effects of impoverished rearing has produced contradictory results. Specifically, rhesus monkey groups (*Macaca
mulatta*) studied at the Wisconsin Regional Primate Research
Center at the University of Wisconsin (Harlow, Schiltz, and
Harlow 1969; Harlow et al. 1971; Rumbaugh and Gill 1976) appear
to possess near-equivalent learning abilities despite radically different early environments. Restricted early environments produce

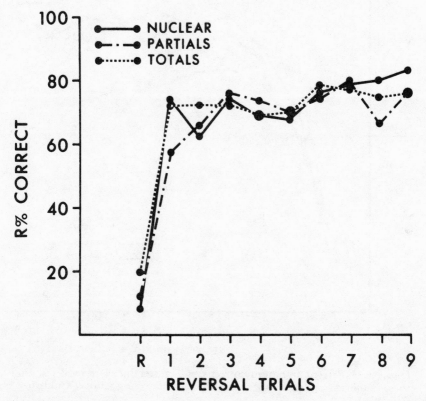

Figure 6.10 Transfer Index values for total and semisocial isolate and for nuclear-family-born-and-reared rhesus monkeys (Rumbaugh and Gill 1976).

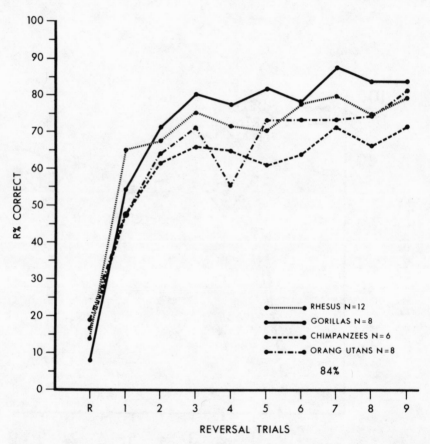

Figure 6.11 Percent responses correct per Transfer Index reversal test trial for rhesus monkeys and great apes. All subjects were young adults (Rumbaugh and Gill 1976).

highly emotional, fearful monkeys; but if these emotional differences are controlled-for through lengthy adaptation procedures
and/or Transfer Index testing, no learning deficits are found
(Figure 6.10), and the performance of these rhesus subjects is again
found commensurate with those of great apes that were wild-born
or raised in standard laboratory environments (Figure 6.11). This is
in marked contrast to the findings that chimpanzees (*Pan*) raised in
restricted environments are significantly inferior to feral-born
chimpanzees (Davenport, Rogers, and Rumbaugh 1973; Davenport
and Rogers 1970) (Figure 6.12). The restricted-reared chimpanzees

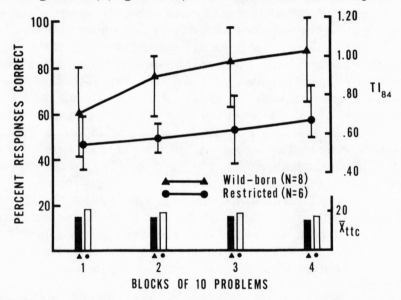

Figure 6.12 Mean and range percentage of correct responses on reversal test trials
2– 10, *left axis*; mean and range of Transfer Index values, *upper right axis*; and
mean trials to criterion on prereversal training, *lower right axis* (Davenport,
Rogers, and Rumbaugh 1973).

appeared more distractible (Davenport and Rogers 1968), relatively
less adaptable and slower to learn (Davenport, Rogers, and Menzel
1969; Rogers and Davenport 1971), possessed inferior tool-using
skills (Menzel, Davenport, and Rogers 1970), and demonstrated
significantly inferior learning skills on the same type of tasks used
to test for cognitive deficits in isolate-reared rhesus monkeys
(Davenport and Rogers 1968; Davenport, Rogers, and Menzel 1969;

Davenport, Rogers, and Rumbaugh 1973). Although many malad-aptive behaviors such as stereotyped movements, excessive fear responses, and inability of isolate-reared subjects to copulate are common to both restricted-reared rhesus monkeys and chimpan-zees, only the isolate-reared chimpanzees, once adapted to the testing situation, show learning deficits.

Since isolate-reared subjects of either species frequently have difficulty adapting to the test situation, the Transfer Index meth-odology is useful as a control for differences in emotionality and distractibility via criterional training. These results indicate that chimpanzee subjects, in contrast to their rhesus counterparts, suffer long-term cognitive deficits as a result of an impoverished early environment. The restricted-reared subjects in Davenport's studies were raised in gray, enclosed cribs from the first day of life to about two years of age. They were fed and diapered via tubes and white mittens inserted through one wall of their crib. During the third year, the restricted environments were gradually enriched and a series of behavioral tests were begun. At the end of the third year, isolation was ended and the restricted-reared subjects were housed in the standard laboratory quarters which also housed the control group. Each group was then equally free to interact socially with other animals. The chimpanzee subjects were adults at the time of TI testing; cognitive deficits attributable to the restrictive early environment persisted even after twelve years of environmental enrichment. Clearly, in chimpanzees, early experience has a pro-found differential effect on the course of future cognitive de-velopment.

Attempts at rehabilitation of deficits accrued by restricted rearing have had varying success. Davenport and his co-workers found the social abberations of the restricted-reared chimpanzees highly resistant to modification by a variety of techniques, includ-ing contact with normal social partners of the same or similar age or administration of various drugs. After several years, instances of play, grooming, and social-sexual contact had increased, and fearfulness and timidity had lessened; but the experimental sub-jects clearly remained deviant in that they continued frequent display of stereotyped behaviors and exhibited few species-specific behaviors (Turner, Davenport, and Rogers 1969; Davenport and Rogers 1970b). In contrast, studies at Wisconsin (Suomi, Harlow, and Novack 1974; Suomi and Harlow 1972) have demonstrated that substantial, if not total, rehabilitation of isolate-reared monkeys can be achieved when rhesus subjects isolate-reared from the first six or twelve months of life are paired with three-month-old socially normal rhesus "therapists" for about eight hours per week. By the end of the six-month therapy period, isolates and therapists did not differ significantly in levels of play, social contact, exploration, or locomotion. Instances of self-clasping, rocking, and stereotype were low for both groups. As the isolates matured, they developed normal sexual patterns and formed well-established dominance hierarchies. These successful recoveries have laid to rest the assumption that damage due to deprived early environments is always irreversible. Unfortunately, the apes in Davenport's studies were exposed to therapists of ages similar to the subjects, and we have no way of knowing if the introduction of less mature animals would have met with greater success. The question as to why younger therapists may succeeed where peers have failed is a fascinating one.

Animals reared in total social isolation are deprived of the periods of prolonged physical interaction in which normal infants *play*. New behaviors, including sex, aggression, fear, and inquisitiveness are gradually incorporated into an infant's repertoire. The infant reared in isolation is denied the opportunity to acquire these behaviors gradually before other, maladaptive behaviors have been incorporated into its repertoire:

> Born with the reflex of clinging and having no mother, surrogate, or peers with whom to cling, the isolates learn to cling to themselves and thus develop self-clasping behavior. Born with the reflex of sucking but denied a maternal nipple for nutritive or nonnutritive contact, they learn to suck their own digits and thus develop high levels of self-orality. Responsive to motor feedback but lacking a source of mobile stimulation, these monkeys provide their own motion and thus develop stereotypic activity (Suomi, Harlow, and Novak 1974).

The isolate subjects, as infants, appear to have learned behaviors which are incompatible with many normal species-specific behaviors; hence these species-specific behaviors do not develop.

If the isolation syndrome is viewed as a learning deficit, it is not surprising that normal peers were not successful therapists—

the normals would be enormously sophisticated in all of those behaviors that the isolate had yet to experience. In the three-month old monkeys, however, play patterns are just beginning to emerge; they still cling to others, and aggression has yet to emerge. The Suomi, Harlow, and Novak isolates developed their social repertoire along with their therapist. When given an infant therapist, the isolate has a social partner with whom all of the manipulative subroutines underlying complex emotions and tasks can be practiced. The two are free to clumsily *play*, each being equally inept (hence less threatening) in their initial social contacts.

OBSERVATIONAL LEARNING AND TOOL USE

By enjoying a prolonged period of immaturity, primate juveniles experience a relatively lengthy period of behavioral plasticity during which conditions are optimal for observational learning. Young primates raised in social groups are more apt to attend to each other and to respond in an imitative or innovative fashion than are adults in the same group (Sackett and Ruppenthal 1973). Also, more rigid social structures of some primate groups apparently preclude imitation of behaviors exhibited by individuals lower in the hierarchy. Since imitation is necessary for observational learning, such a social structure allows for juveniles to observe and learn from one another or from their mothers, but precludes adult imitation of juvenile behavioral innovations.

Perhaps the two most famous examples of the fruits of youthful plasticity being rigidly confined by the social structure stem from a female monkey (Imo, a *Macaca fuscada*) on Koshima Island (Kawai 1965). The monkeys on this island are provided with potatoes and wheat which are regularly dumped on the sand. One day, when Imo was two years old, she was observed washing a potato in a stream, thus removing the clinging sand before eating it. In time, Imo's age-mates also began washing their potatoes, but few adults were seen doing so. But when Imo and her peers had infants of their own, these infants, who had observed their mothers' washing behaviors, also began washing their potatoes.

Imo also instituted a means for separating wheat from sand. When she was four years old, she was observed throwing a handful of wheat and sand into the water. The sand sank while the wheat floated where it could be picked, now clean, from the water. This practice of throwing food *away* to clean it spread among Imo's age-mates and to some younger members of the troop. The older monkeys either retained their old method of picking the grain from the sand or else stood downstream from a grain washer and picked out bits of wheat floating past. In each of these instances, the "discovery" was made by a young individual presumably still relatively plastic in her range of potential behaviors and was transmitted to others who were still capable of observing and imitating new behaviors.

Chimpanzees do not appear to be so restrictive about acquiring new behaviors by observing others lower in the social hierarchy, and they also seem innovative further into adulthood than macaques. Although some behaviors such as termiting are apparently passed exclusively from mother to child, many forms of observational learning appear less constrained. This may be due in part to a more relaxed social structure permitting more observational learning (for a discussion, see Bruner 1972). Also, instances of chimpanzee tool use are much more widespread and complex than those seen in monkeys (Beck 1974). Loosening of the social structure, the prolongation of immaturity, and the increase in cortical development seen in apes all favor emergence of observational learning as a primary, not fortuitous, means of learning. Physical characteristics of apes and monkeys are very similar, yet apes' manipulative skills are vastly superior. This has been experimentally demonstrated (Parker 1974) but is also verifiable by "neofield observation": walk through any primate area where both monkeys and apes are housed and note the relatively elaborate strongholds erected to contain apes. Yet apes rarely escape by brute force; they manipulate by picking and pulling and turning and knocking until the screws fall away.

Parker (1974) obtained data concerning the various types and complexity of manipulatory behaviors observed by subjects representing six major taxonomic divisions. He used one pair each of ring-tailed lemurs (*Lemur catta*) and black lemurs (*Lemur macaco*) to represent the prosimians; spider monkeys (*Atteles geoffroyi*) and capuchins (*Cebus capuchin*) to represent the Ceboids; silvered leaf-eating langurs (*Presbytis cristatus*) for Cercopithecoids and Colobinae; mona guenons (*Cercopithecus mona*) and pig-tailed macaques (*Macaca nemistrina*) for Cercopithecoids; white-handed gibbons (*Hylobates lar*) for Hylobatidae; and the common chimpanzee (*Pan troglodytes*), lowland gorilla (*Gorilla gorilla*), and orangutan (*Pongo pygmaeus*) represented Pongidae. Each subject was isolated in a test cage into which one end of a nylon mountaineering rope protruded. The subject's interactions with the rope were observed and classified into various primary or secondary action categories. Primary actions were those in which the manipulandum was directly acted upon by the body part in contact with it (e.g., touching or biting the rope, walking with it, pulling apart the rope strands), whereas secondary actions were those in which the manipulandum was applied, or used in relation to an object (e.g., winding the rope around the body, rubbing another body part with the rope, striking an object with the rope). Great apes exhibited more behaviors in more diverse categories than did any other group. The pongids also displayed significantly more secondary actions than did any other group. Table 6.3 shows the species tested arranged in descending order on the basis of mean percent of secondary actions. Since secondary actions are necessary for tool use, these data are as would be predicted, based upon widespread reports of great ape tool use and relatively rare instances of tool use for the other nonhuman primates.

Wild chimpanzees are known to make and use tools with far

Table 6.3 Species in Rank Order for Mean Percent of
Behavior Classified as Secondary*

Species	%
Orangutan	35.59
Chimpanzee	28.73
Gorilla	12.81
Capuchin	6.25
Macaque	4.85
Gibbon	3.54
Lemur	2.84
Langur	2.05
Guenon	0.00
Spider Monkey	0.00

*The vertical lines indicate the results of Duncan's New Multiple Range Test. Species not joined by the same line are significantly different from each other at the .05 level (*left*) or the .01 level (*right*). (Parker 1974).

greater frequency and variety than wild baboons. Beck (1974) has noted that, although fortuitous discovery of tools may be equally rare for both chimpanzees and baboons, the new information, once discovered, is transmitted quickly among chimpanzee groups by observational learning and soon becomes well established in the group's repertoire. In contrast, baboons acquire little information by observation. For baboons, observation of another baboon using a hooked stick to rake in food *might* increase the frequency of orienting toward the food while holding the stick, indicating that some information about the task had been transferred by observation; but this stimulus enhancement may have resulted from the tool's acquisition of secondary reinforcing properties due to its previous association with food.

Information transmission can occur very quickly in chimpanzee groups. Menzel (1970, 1973) has observed confined chimpanzees using poles to escape to a platform booth above the large, walled-in area. The initial break-in was at night and went unobserved; but next morning, most of the chimpanzees were seen propping poles against the wall and climbing them to gain access to the booth. On the day after the initial break-in, five out of eight chimpanzees had mastered the propping technique. Two months later, one chimpanzee propped a branch in such a way as to form a ladder to treetops outside the enclosure. Within a day of this discovery, all animals were using branch ladders for escape to the treetops. What is impressive and indicative of observational learning here is the speed with which this new behavior was simultaneously acquired by several troop members. Branches had been accessible to the animals for years before the single discovery which was observed by other troop members and then immediately replicated.

Gombe Stream chimpanzees and baboons inhabit the same area, thus have access to the same implements for fortuitous tool discovery. Yet only the chimpanzees have been observed to use many different types of tools. Even if the chimpanzees' behavioral repertoire favors fortuitous discovery of tools more than the baboons', the baboons do not appear to possess observational learning skills sufficient to learn from the chimpanzees. Chimpanzees throw rocks at baboons. Baboons see/feel the rocks, but throw nothing at chimpanzees (van Lawick-Goodall 1968).

LANGUAGE ACQUISITION BY NONHUMAN PRIMATES

Anthropoid behaviors exhibited by chimpanzees have led various researchers to speculate that chimpanzees may be artificially induced to acquire language. The relatively extensive repertoire of vocal calls and gestures exhibited by free-ranging chimpanzees certainly serves to communicate information (van Lawick-Goodall 1968), but such communication is not generally considered to fall within the realm of language. Although no undisputed definition of language exists, most definitions would preclude chimpanzee communication in the field because of its (apparent) lack of abstract symbols or grammatical structure. Human language, unlike the communication systems of nonhuman forms, combines initially *arbitrary* symbols in various ways as an attempt to communicate. This facility for usage of arbitrary symbols is additional to the more apelike communication systems, in which calls or gestures are either elicited by, or imitative of, the information to be conveyed.

Neurological basis for the anatomical sophistication necessary may be the same as for many complex cognitive skills demonstrated by both humans and apes. The human condition in contrast to the ape, however, is such that language skills normally develop. If the potential for language is not uniquely human but shared by other species with suitably evolved nervous systems, such species should exhibit language when placed in a suitable environment. Clearly, the natural environment of the chimpanzee is not supportive of language development; but this does not preclude such development in a different, more language-conducive surround. Even humans cannot swim unless given an appropriate medium; similarly, environmental constraints appear to dictate the language potential of chimpanzees.

Evidently, raising an ape in a human home, as a child would be raised, is not sufficient for language acquisition sufficient for two-way communication. Hayes and Hayes (1952) trained their home-raised chimpanzee Vicki by manually shaping her mouth into appropriate position for emitting human speech. After six-and-a-half years of training, Vicki could mouth *unvoiced* whispers of "mama," "papa," "cup," and "up." Although Vicki could respond appropriately to many spoken words, she was unable to convey such information via human speech. In all probability, the chimpanzee vocal apparatus is incapable of producing the phonemes used in human speech (Lieberman, Crelin, and Klatt 1972). But a language need not be spoken; it can be read or conveyed by hand signals such as those used in the American Sign Language for the deaf (ASL). Attempts at teaching language to a chimpanzee by these means have met with great success.

The Gardners (Gardner and Gardner 1969, 1971) taught their chimpanzee Washoe more than 130 ASL signs/words, some of which Washoe spontaneously strung together in ways appropriate to their use in novel situations. For instance, Washoe signed "key open please blanket" to have a locked blanket cupboard opened. Fouts (1974) has continued to work with Washoe and has instituted ASL training with other chimpanzees. These chimpanzees have been observed to communicate with each other using ASL in situations involving play, food sharing, and mutual comforting (Fouts, Mellgren, and Lemmon 1973). They have repeatedly demonstrated that they can recombine signs in their vocabulary to describe new objects or situations. Washoe has called ducks "water birds." Another chimpanzee, Lucy, signed "cry hurt food" after tasting a radish.

The order of Washoe's signing suggested what may have been rudimentary syntax—rules for sentence structure—in changing the meaning of a string of signs by rearranging the order of the signs. Studies by Premack (1971) extended justification for the claim that chimpanzees could demonstrate syntax. To designate words, Premack used plastic pieces that could be arranged on a magnetic board. His chimpanzee Sarah would respond appropriately to commands such as "Sarah insert apple pail banana dish" by first placing the apple in the pail, then placing the banana in the dish. Sarah could also use the plastic symbols to describe attributes of objects not present at the time. In response to the blue chip representing "apple," Sarah selected symbols for "red," "round," and "with a stem," and responded "not" to "green," "square," and "without a stem." Clearly, the blue chip arbitrarily designated to signify "apple" had come to represent the attributes of the apple rather than those of the chip itself.

The chimpanzee's speech limitations have been circumvented

in yet another way. A computer-controlled language training system is currently in use at the Yerkes Regional Primate Center (Rumbaugh et al. 1973). The chimpanzee subject, Lana, lives in a seven foot Plexiglas cube which allows her twenty-four-hour-a-day access to a keyboard which currently holds seventy-five keys. Embossed on each key is a geometric figure (lexigram) which represents a specific word of a word group. Background color of the key designates the class of words to which the lexigram belongs—animate beings, physical objects, verbs, foods, prepositions, etc. As keys are depressed, lexigrams on the key faces appear on a row of projectors directly above Lana's keyboard (Figure 6.13)

Figure 6.13 Lana, age three-and-a-half years, at the keyboard console of her computer-controlled language training system (photo courtesy of Timothy V. Gill).

and on a series of projectors outside her room above the experimenter's keyboard. The latter keyboard is the functional equivalent of Lana's, but has no lexigrams; instead, buttons are labeled with the English equivalent of the lexigrams. Key depressions are continu-

ously monitored by a PDP 8 computer to record all that transpires
and to determine if the phrases constructed are syntactically correct
with respect to rules of the language (called *Yerkish*) used. Since
training began, in January 1973, Lana has been required to attend
both to the meaning of specific lexigrams and to their appropriate
place in a sentence. The computer responds to appropriate requests
by generating signals to open a window, dispense various foods or
a toy, play music, or show slides. If a human operator is present,
this system also allows for conversation between human and ape.

Lana came to attend to the projected lexigrams on her own
and was sensitive to the order in which the lexigrams appeared. She
would successfully complete sentences that were correctly begun by
an experimenter—for example, she would add "juice" or "M&M"
to sentences beginning "Please machine give"—but she would
erase incorrect beginnings such as "Please give machine" by
depressing the "period" key (Rumbaugh, Gill, and von Glasersfeld
1973).

Lana has also learned that objects have names and that a given
name may represent several distinct objects (Gill and Rumbaugh
1974). Naming training began by showing Lana either M&M
candies or pieces of banana. She was required to respond to "? what
name-of this period" with either "banana name-of this period" or
"M&M name-of this period." This initial training proceeded
surprisingly slowly, and about 1,600 trials were required before
Lana reliably named these two incentives. But less than five trials
taught Lana the names of the next five incentives introduced.
"Milk" and "blanket" were correctly named on their first presentation. This rapid rise in performance is characteristic of what is
generally termed "insightful" learning. It indicated that Lana had
acquired the general concept that things have arbitrary names, and
she soon began applying names to objects not specifically
requested. When a slide projector was installed in Lana's environment, an experimenter typed through his keyboard "Please
machine make slide period," at which point the computer activated the slide projector and displayed three slides for fifteen
seconds each. When the slides had finished, Lana immediately
went to her keyboard and typed "slide name-of this period." Since

this time, Lana has begun asking for names of previously unnamed objects and then using the new names correctly. The following conversation occurred when an experimenter, sitting just outside of Lana's room, held in his hand a cardboard box baited with M&M candies in full view of Lana. The key for "box" on her keyboard had not been previously used.

May 6, 1974.

Lana: *? Tim give Lana this can.*   11:36 A.M.

Tim: *Yes.* (And Tim gave her an empty can, though apparently she wanted the box with the M&M's.)

Lana: *?Tim give Lana this can.*   11:42 A.M.

Tim: *No can.* (Which meant that Tim did not have the can, as he had just given it to her.)

Lana: *? Tim give Lana this bowl.*   11:43 A.M.

Tim: *Yes.* (And Tim gave the empty bowl to her.)

Lana: *? Shelley* (sentence unfinished).   11:43 A.M.

Tim: *No Shelley.* (Shelley, a technician, was not present.)

Lana: *? Tim give Lana this bowl.* (Before Tim could   11:44 A.M.
answer, she continued . . .)

Lana: *? Tim give Lana name-of this.*   11:45 A.M.

Tim: *Box name-of this.* (Tim thereby gave her the name of the vessel which she apparently wanted.)

Lana: *Yes.*   11:46 A.M.

Lana: *? Tim give Lana this box.*   11:47 A.M.

Tim: *Yes.* (Whereupon Tim gave it to her, and she immediately ripped it open and extracted the M&M candies.)

Repetitions, in essence, of these proceedings took place later the same day when Lana wanted a cup that contained M&M's. She first asked Tim for its name, then asked Tim for it by name—*cup*. (Rumbaugh and Gill 1974).

Lana had never been taught to ask for the name of an object. That she did so indicates that she imitated a technician's behavior to gain a similar, though not identical, end. When a technician asked for the name of an object, it had always been within a testing situation where the object name had already been displayed at least once. In this situation, Lana asked for the name of a previously *un*named object.

Washoe, when she saw some Brazil nuts for which she had no specific sign, asked for the "rock berry" (Fouts 1974). As Lana's keyboard facility (which we believe correlates highly with her conceptual facility) increases, there is a corresponding increase in instances of novel word-and-phrase combinations to describe or comment upon previously unmentioned events. The following conversation took place when Lana had been in the language program for about seventeen months. It commenced when Tim Gill entered the anteroom with a sliced orange, one of Lana's preferred fruits for which she had no name:

> May 28, 1974.
> Tim: *? What color of this.*
> Lana: *Color of this orange.*
> Tim: *Yes.*
> Lana: *? Tim give cup which-is red.*
> Tim: *Yes.* (And he gave her a red cup, which she discarded, and continued—)
> Lana: *? Tim give apple which-is green.* (Lana occasionally confused the keys for the colors orange and green at this time.)
> Tim: *No apple which-is green.* (A way of saying, "I have no green apple to give.")
> Lana: *? Tim give apple which-is orange.* (Whereupon she bounded with suggested enthusiasm to the door to receive "the orange-colored apple.")
> Tim: *Yes.* (And he gave it to her.)

When faced with less-than-perfect performance of seemingly simple tasks, experimenters have wondered if tasks such as those used to measure learning set might bore an ape. These apprehensions appear well-warranted. For, although the discrimination testing situation may well form the most interesting event in a captive ape's day, the task per se is probably trivial. As new and more demanding research strategies are devised and employed, even the currently impressive sayings of "rock berry" and "apple which-is orange" are sure to sound truly infantile. But recognition that we have not yet fully assessed the cognitive capabilities of apes was necessary before what would otherwise have been considered fantastic research could proceed. Such research may well contain information necessary for the survival of anthropoid, as well as of other, forms of life.

7

Basic Operant Research in the Zoo

*Victor J. Stevens**

For many people, the term "operant conditioning" brings to mind an image of a small box in which a starving white rat frantically presses a bar for food pellets. Although this stereotype is not fair to the entire field of the experimental analysis of behavior, it is not a particularly inaccurate caricature of much research in this area. "Skinner box" research using a few select species of domestic animals seems to be far removed from anything natural and an unlikely place to look for basic principles of behavior. Yet the expressed goal of operant conditioners is to develop principles which are generalizable to the behavior of all organisms in virtually any setting. This chapter reports a series of operant research projects with zoo animals and compares them to analogous laboratory studies with special attention to the question of generality.

Applying Laboratory Techniques to Zoo Research

Operant researchers most commonly use standard strains of laboratory-bred Norway rats or domestic pigeons as subjects. Such animals are typically housed in small cages in a laboratory,

*At the time this chapter was written, Victor J. Stevens was visiting assistant professor of psychology at Reed College, Portland, Oregon. He is currently an associate scientist and director of research education at the Oregon Zoological Research Center, Portland Zoological Gardens.

reduced to 80 percent of their free feeding weight, then placed individually in small, soundproofed boxes for each experimental session. In most cases, rats press bars and are reinforced with food pellets and pigeons peck small plastic disks (keys) and are reinforced with a few seconds' access to a grain bin. Generally the interiors of these boxes are very plain, the only prominent features being response levers or keys, a food cup, a house light, and perhaps a stimulus light. Numerous variations on this setting exist, of course, with new modifications appearing regularly. Still, the basic Skinner box rarely deviates far from the situation described, because the degree of environmental control it provides has resulted in very consistent data.

Since most operant researchers do not stress species differences but search for basic similarities among individuals and species, experimental procedures which result in unexplained variability are modified until the controlling variables for these effects can be determined. Procedures which lead to broadly applicable principles, both across species and situational differences, become the methods of choice. Therefore, to test generality and often for other reasons as well,[1] the operant box has been modified for use with increasing numbers of animals including porpoises (Turner and Norris 1966), bats (Beecher 1971), owls (Martin 1974), Siamese fighting fish (Thompson 1963), turtles (Bitterman 1964), and honeybees (Grossman 1973). Highly consistent results have been found in dealing with a large number of variables including: reinforcement schedules, discrimination performance, choice between concurrently available responses and schedules, effects of conditioned reinforcers, prediction of the power of a reinforcer, and many others. Relatively few tests have been conducted, however, of the generality of the effects of such variables in the less controlled, nonlaboratory situations (with the exception of human behavior therapy research[2]).

---

1. Many operant studies using atypical subjects have been designed to study special characteristics of a particular species including sensory capacities, navigational skills, memory, comparative reinforcer magnitudes, strength of imprinting, and so on.

2. For recent reviews of behavior therapy research see Bandura (1969) or Rimm and Masters (1974).

Operant research in the zoological park offers fairly challenging tests of the generality of laboratory results for several reasons (Markowitz and Stevens 1974). First, a variety of species are available, most of which are very dissimilar to "standard" laboratory animals. Second, the soundproofed, light attenuated operant boxes of the laboratory are unsuitable since zoo administrators are understandably reluctant to remove animals from display, even for short periods. (Of course, sound-attenuating laboratory facilities would not usually be available anyway, since every zoo operates on a limited budget.) Third, because of health considerations and appearance of the animals, zoos do not favor allowing reduction of individuals to 80 percent of their free feeding weight. Since most species are fed once daily, however, a mild form of food deprivation is already achieved.[3] Fourth, because many modern zoos are actively engaged in breeding programs for exotic and endangered species, any procedures which could conceivably interfere with reproduction are prohibited. For this reason, curators may object to animals being separated from each other even for short periods.

All of these factors lead to a situation in which the operant researcher is working with relatively unknown exotic species in noisy, fairly uncontrolled settings while the animals are exhibited in social groups. Findings that diverged from typical laboratory-based research would be of limited value in the experimental analysis of behavior because of the many variables that could account for differences. These problems may have discouraged operant researchers in the past. But, on the brighter side, any similarities between zoo and laboratory findings would be an impressive demonstration of generality.

The remainder of this chapter will report the findings of a series of basic operant research projects conducted at the Portland Zoological Gardens, Portland, Oregon. These findings illustrate the usefulness of the zoo setting in conducting operant research

---

3. The measurement of food deprivation in terms of free feeding body weight tends to obscure the fact that most free living wild animals are well below this figure much of the time. Since in practice most zoo animals are also kept below their free feeding weight for health and appearance reasons, researchers can effectively use food as a reinforcer. If further deprivation is needed, however, it must be done carefully, especially in the case of ungulates (Wenzel, Baldwin, and Tschirgi, 1960, and Squier, 1964).

under conditions much different from the standard laboratory setting. They also add to a growing body of data showing the generality and utility of behavioral principles dealing with wild animals in both social and nonsocial settings.

ELEPHANTS EXPOSED TO FIXED RATIO SCHEDULES AND DISCRIMINATION TRAINING

The first basic operant research carried out in a zoo was Leslie Squier's 1964 work with the Portland Zoo's famous herd of Indian elephants (*Elephas maximus*).[4] Squier's first problem was selection of a suitable reinforcer since he was unable to deprive his subjects of food or dispense a significant proportion of the regular hay diet. Two gram sugar cubes were finally chosen. Surprisingly, this highly preferred treat is a very effective reinforcer for these huge creatures and can be used to maintain behaviors involving substantial expenditures of energy.

Because captive elephants tend to destroy anything they can get their trunks on, Squier also had considerable difficulty in developing an appropriate response station. Heavy-duty chains and response levers were torn from their mounts, for example, during the subjects' casual inspection of the equipment. Finally, a large, heavy plywood box was built and positioned outside the cage area so that the subjects could just touch the front panel with the end of their trunks. The response consisted of pushing a thirteen-centimeter Plexiglas panel, and the sugar cubes were dispensed into the manger below. Since the animals could not be separated or removed from their exhibits, they were often run in groups. Dominance hierarchies quickly became apparent; frequently one animal would drive the others away, and then only one elephant in the group could be used as a research subject.

During the course of the research, one adult male and four adult females were used as subjects. Each was trained to key press in less than one session. The speed of acquisition can be seen in Figure 7.1, which shows the cumulative records of two adult females on their first day of training. A juvenile male and an infant

4. This research was conducted by Leslie Squier from the summer of 1963 to the fall of 1964. Unfortunately most of the data were lost in a fire and the records reproduced here are the only ones that survived.

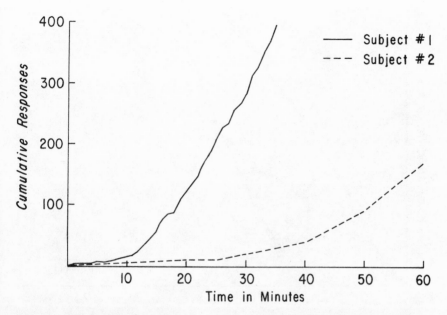

Figure 7.1 Cumulative response records for two adult female elephants during their first key-pressing sessions.

female were also trained to respond, but each required several sessions to learn the task. When exposed to variable ratio schedules, in which reinforcement follows a variable number of responses, the subjects all showed the rapid response rates typical of those generated in the laboratory (see Figure 7.2). (Ratio schedules will be discussed in more detail in the following section.)

Several adult females were eventually exposed to response requirements as high as 50 without showing any signs of ratio strain (i.e., irregular bursts of responding and long pauses). At high ratio values, they showed sustained overall response rates of approximately 67 per minute; for one female, the rate went as high as 100 per minute. Each individual's response rate tended to increase with increasing response requirements, as is commonly seen in laboratory research.

In a later portion of the study, three females were trained on a simple light-dark discrimination (i.e., in a choice situation, responses on the lighted panel were reinforced and responses on

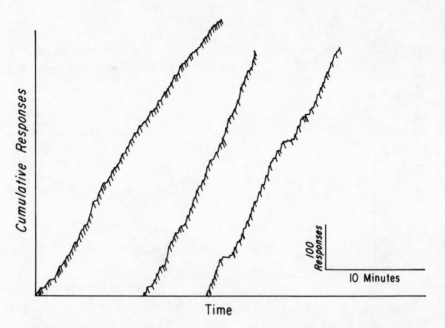

Figure 7.2 Cumulative response record of an adult female elephant on a variable-ratio 10 schedule of reinforcement. (Diagonal marks indicate reinforcer delivery.)

the dark key were not reinforced). They also quickly learned this task and were responding correctly more than 90 percent of the time after only a few sessions. Shortly after this training, the research was terminated (due to adverse administrative policies in the zoo). When the research was resumed eight years later, the one subject who had not developed substantial visual impairment responded as if no delay had occurred (Markowitz et al. 1975). She made only two errors and had reached a criterion of twenty consecutive correct responses within the first 6 minutes. The elephant's reputation for good memory seems well deserved.

Regarding elephants as research subjects, Squier reported that aside from equipment problems, they were easy to work with, quick to respond to contingency changes, and were very affectionate. They seemed to enjoy the experimental sessions and were not affected by most extraneous stimuli such as noises made by visitors or passing trucks. In short, in many ways he found them to be ideal subjects for basic operant research.

CAMELS EXPOSED TO FIXED-RATIO SCHEDULES

Operant research with dromedary camels (*Camelus dromedarius*) began in the spring of 1973.[5] We wanted to develop research techniques suitable for ungulates in the zoo setting, and camels seemed an ideal species to start with because of their size, apparent serenity, and large, undiscriminating appetite. Also, we were interested in learning whether they could really be as ornery and stupid as their reputation would indicate. Since we found no mention of previous learning research using camels, we developed techniques as we went along.

At the outset of research, the Portland Zoo collection included three adult females and one adult male.[6] We used all four of them as subjects over the course of 2 years, plus two calves born during that period. The camels were housed in one of two display enclosures in groups as large as four. Since normal zoo routine made it impossible to separate individuals for experimentation, the research was conducted while they were on display in their home enclosures.

We selected "universal ungulate pellets" for a reinforcer. This highly preferred food, a part of the regular zoo diet, was sufficiently powerful to be effective immediately following scheduled feedings and also when nonpreferred food (such as hay) was present. Early experimentation showed that as few as 15 grams of these pellets would induce the camels to cross their pen (about twenty meters).

An automated feeder to dispense the pellets was fabricated from used vending-machine parts. On the advice of the animal keepers, we mounted it behind a heavy plywood panel attached to a crib inside the camel barn. The side of the crib was enclosed so that only one animal at a time could gain access to the feeder by

---

5. Conduct of this research included the aid of Georgianne Schmuckal, Hal Markowitz, Tom Cox, and Janet Baldwin. Jan took care of most day-to-day chores including running animals, repairing equipment, and organizing volunteers. Some of this research has been reported elsewhere: Markowitz and Stevens (1974), Baldwin, Stevens and Markowitz (1973), Stevens et al. (1973), and Markowitz and Stevens (1974).

6. These animals were trapped as juveniles from feral populations in Australian deserts. Australian camels are preferred by zoos because Australia has no hoof and mouth disease and quarantine procedures are simpler.

stretching its neck around the corner. We felt these measures necessary, at the time, to protect our equipment from aggressive camels, but experience proved such precautions quite unnecessary. In view of their size and strength, the camels have treated the research equipment quite gently.

The feeder made a very loud, sharp noise during its operation; but much to our surprise, the camels habituated to this sound faster than did the experimenters, and the animals were magazine-trained (i.e., trained to come from the other side of the enclosure following feeder operation) in just one session. The male soon began keeping the females away from the feeder during the experimental sessions and, for this reason, became the primary subject for the first part of the study. The females could only be used when the male had become food-satiated and left the immediate area. When, after a few months, the male was moved to an adjacent enclosure, the dominant female became the primary subject.

For a measurable operant response, we started with eight by twelve-centimeter Plexiglas panels that required about 50 grams force to close a microswitch. This type of response panel had been successfully used with elephants (Squier 1964), goats (Wenzel, Baldwin, and Tschirgi 1960), and harbor seals (Markowitz 1975); and we assumed that camels, with their prominent noses, could easily be trained to nudge these panels. In two years of work, however, none of the subjects responded in this manner. Instead, the four camels trained all pushed it with their lower jaw incisors. This response was easily shaped in one or two sessions by systematically reinforcing successive approximations. The male was shaped, for example, by an inexperienced research assistant in two sessions after only sixty reinforcers. Typically, individuals would not show signs of satiation until receiving at least fifty 15-gram reinforcement deliveries, and they would often continue to respond after receiving two hundred or more in one session.

Following initial shaping, the male was run in a series of 1.5 hour daily sessions in which the number of responses required before reinforcement (fixed-ratio value) was gradually increased. Each fixed-ratio (FR) value was held constant for a few sessions until responding stabilized. In this way, the male was slowly worked up to an FR value of 150.

After several months, we became concerned about the camel

tooth wear that might occur from pushing of the panel with their incisors, so we installed a touch switch activated by the animal's body capacitance.[7] For a response, the animal had simply to touch a four by five centimeter stainless steel plate with an electrically conductive part of the head (for example, the lips, tongue, or tip of nose). This response was also quickly shaped, the animals usually responding with their lips. Using the touch switch, the male camel was extended to an FR 20 before he was moved to another cage for breeding purposes. Female 1 was then exposed to a series of FR values from 1 to 35.

Ratio values were controlled, and responses were recorded with conventional relay equipment placed near the response panels. The click from a solenoid in the cumulative recorder provided feedback for a correct response.

Analysis of cumulative records and observation of the subjects during experimental sessions showed that the camels' behavior was similar in all respects to that of other species responding to FR schedules. The typical response patterns of rats, pigeons, and humans on FR schedules in the laboratory setting consist of pauses after reinforcement, followed by an uninterrupted burst of steady, rapid responding that terminates in reinforcement (Ferster and Skinner 1957). Cumulative records of performance following several sessions' exposure to a particular schedule clearly show that our camels responded in the typical manner for both the push response (Figures 7.3 and 7.4) and the touch response (Figure 7.5). The only major difference between our records and those collected in laboratory studies is an occasional long break in responding. These anomalies, fairly common in our zoo research, are usually attributable to major distractions such as jet planes, zoo visitors throwing food into the exhibit, or passing trucks. As research progressed and the subjects became more experienced, the frequency of these breaks decreased from an average of several times a session to less than once per session.

Our inability to maintain responding at ratio values greater than 150 was probably a function of the complex situation. Specifically, other sources of food in the environment, such as hay and food begged from zoo visitors, were often available. These options were concurrent schedules; and in such situations in the

7. Model #202A touch switch, Raven Electronics, Sparks, Nevada.

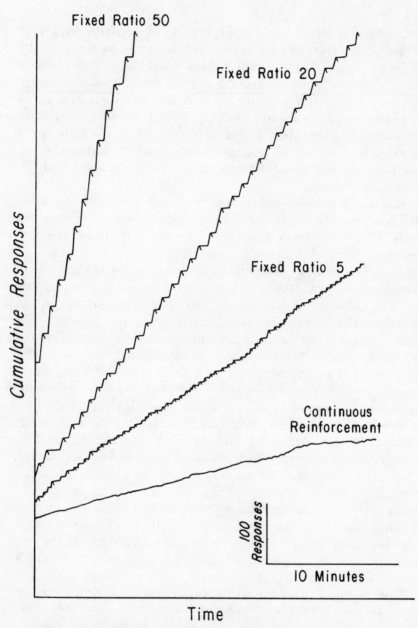

Figure 7.3 Cumulative response records of the stability performance of an adult male camel on fixed-ratio reinforcement schedules (push response).

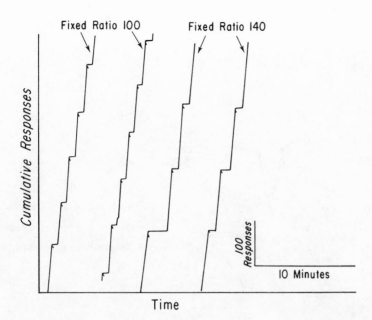

Figure 7.4 Cumulative response records of the stability performance of an adult male camel on fixed-ratio reinforcement schedules (push response).

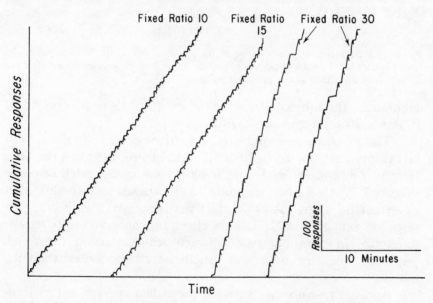

Figure 7.5 Cumulative response records of the stability performance of an adult male camel on fixed-ratio reinforcement schedules (touch response).

219

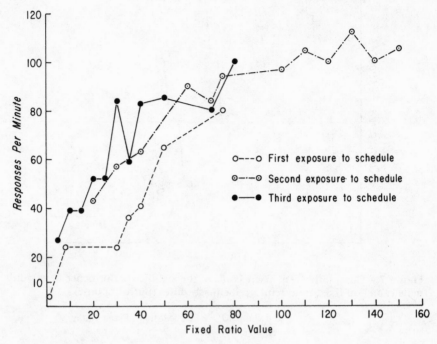

Figure 7.6 Stability response rates of the male camel on fixed-ratio schedules (push response). In each exposure sequence, the FR value was increased after the subject had stabilized at the previous value.

laboratory, the subject selects the richest schedule of reinforcement (Catania 1966; Herrnstein 1970).

The response rate increased with increasing FR values, as in laboratory data, for both the male camel using the push response (Figure 7.6) and the male and female 1 using the touch response (Figure 7.7). Also, when the male was reexposed to a schedule after experiencing larger FR values, he usually showed a slightly higher response rate (Figure 7.7). This effect probably accounts for the difference in rates observed between subjects using the touch response, since the male was much more experienced than the female.

Another resemblance between these data and laboratory findings is the tendency for the post-reinforcement pause (PRP) to increase as a function of FR value. For example, during the last series of schedules on the push response the mean PRP was 13.8

seconds for FR 20, 16.8 seconds for FR 40, 24.6 seconds for FR 70, and 27.6 seconds at FR 90 (Figure 7.8). The mean PRP tended to decrease slightly upon reexposure to an FR value. These findings, while consistent with results of laboratory studies with rats (Premack, Schaeffer, and Hundt 1964) and pigeons (Felton and Lyon 1966), showed, however, a somewhat greater increase in PRP than was exhibited by the camels.

No appreciable change in PRP was observed in either subject when using the touch response. The mean PRP for the male was 12.0 seconds at FR 10 and 14.5 seconds at FR 15 and FR 20. For the female, the mean PRP was 13.5 seconds at FR 10, 13.5 seconds at FR 20, and 13.2 seconds at FR 30 (Figure 7.9). These comparatively short post-reinforcement pauses and small increases in PRP with increasing ratio requirements were seen with both touch and push responses. The slight variance from laboratory findings is probably a function of the relative response cost (i.e., difficulty) associated with the various situations. From the experimenters' view, the touch response seemed much easier for the camels, and the male showed slightly higher response rates when using that response. Use of a more effortful response with camels might well result in longer PRP's. An additional factor which might have contributed to short PRP's was the social setting, in which a long pause may encourage another camel to move in and attempt to respond. In any case, however, the response to FR schedules was strikingly similar to results of data collected from different species in considerably more controlled settings.

It should be added that contrary to their reputation, we found camels very easy to work with. No researcher was ever threatened or attacked, and the animals were never observed to spit at anyone. Typically, they were calm, very responsive to contingency changes, and generally made excellent subjects for operant research. (Perhaps we would have seen another side of their nature had we forced them to carry heavy loads across the desert.)

Research on fixed-ratio schedules was terminated by an unfortunate series of events including equipment breakdowns, staff shortages, and the beginning of the academic year. When research resumed months later, we decided to expose the camels to another condition which had never been studied outside of the laboratory.

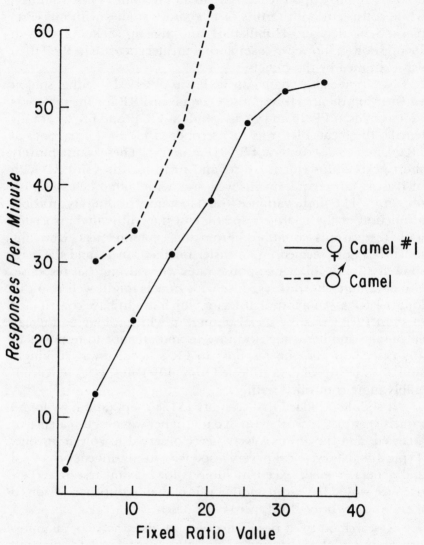

Figure 7.7 Stability response rates for two adult camels on fixed-ratio schedules (touch response). The FR value was increased for both subjects after stability performance had been achieved at the previous value.

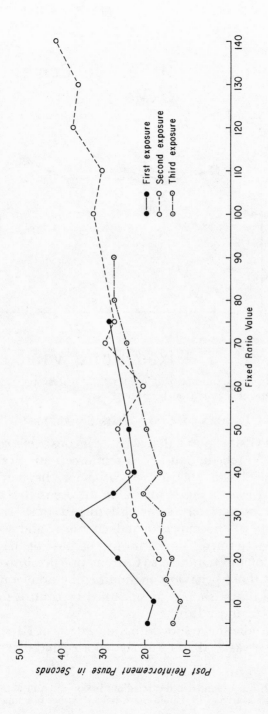

Figure 7.8. Mean post-reinforcement pauses exhibited by the male camel on various fixed-ratio schedules (push response).

223

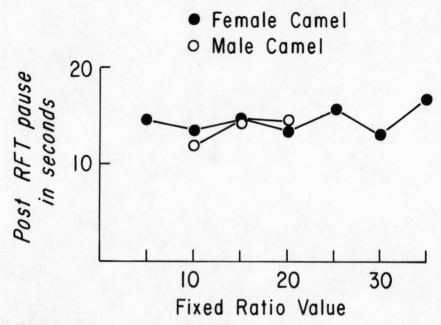

Figure 7.9 Mean post-reinforcement pauses exhibited by two adult camels on various fixed-ratio schedules (touch response).

Camels on Fixed Time Schedules[8]

Skinner (1948) found that when pigeons were exposed to a fixed time (FT) schedule in which reinforcers are given at regular intervals regardless of the subject's behavior, they responded as if their behavior was, in fact, controlling the occurrence of reinforcement. Before food delivery, some birds turned circles, others pecked at one corner of the experimental chamber, and so on. These behaviors were termed "superstitious" because of the noncontingent nature of the reinforcers. Current practice also includes as superstitious those behaviors maintained by noncontingent reinforcers but initially shaped or sustained by contingent reinforcement (e.g., Neuringer 1970).

Recent studies have shown that the effects of FT schedules are much more pervasive than formerly reported and that a number of

8. This research was designed and carried out in conjunction with John Penney, January through March 1975. Linda Stark helped gather baseline data.

behaviors may be significantly affected. Staddon and Simmelhag (1971) exposed pigeons to several schedules of noncontingent reinforcement and systematically observed their behavior throughout each session. They found that "terminal" behaviors (i.e., those reported by Skinner 1948) occurred with high probability at the end of an interval just before reinforcement delivery. Also, "interim" behaviors occurred with the greatest probability midway between reinforcers, and at low probability just before and after reinforcement. Although the specific behaviors varied somewhat between individual birds, this general pattern was clearly seen in each subject.

When we decided to see how FT schedules affected the behavior of camels, the four individuals in the exhibit where our feeder was mounted included two adult females (2 and 3) and their calves (4 and 5). They were first observed for four 1-hour sessions to determine their baseline patterns of behavior. Next, the FT schedule was instituted by programming the feeder to operate every ninety seconds regardless of the subjects' behavior. Both the push-response key and touch-response plate were left in place, but their operation had no effect. In this initial portion of the research, sessions were scheduled to occur a few hours before the regular daily feeding. For the purpose of data collection, behavior was classified into eight mutually exclusive categories including: eating (or licking the food tray), pushing the response panel or touch plate, facing the response key while motionless, facing the equipment box, head bobbing, biting another animal, being between one and five meters from the feeder (and emitting none of the aforementioned behaviors), and being more than five meters from the feeder (or out of the observer's view). The observer recorded the sequence of each subject's behavior in ten-second blocks starting with reinforcer delivery.

As in the fixed-ratio study, the camels rapidly developed a hierarchy and the dominant female 2 kept the other camels away following food deliveries. After female 2 satiated and left the feeder area female 3 would defend that area. An occasional nip and much roaring and open-mouthed threatening occurred during dominance conflicts, but none of the camels were injured during the course of research. The subjects were also observed fighting intermittently at their routine feeding times and when zoo visitors threw

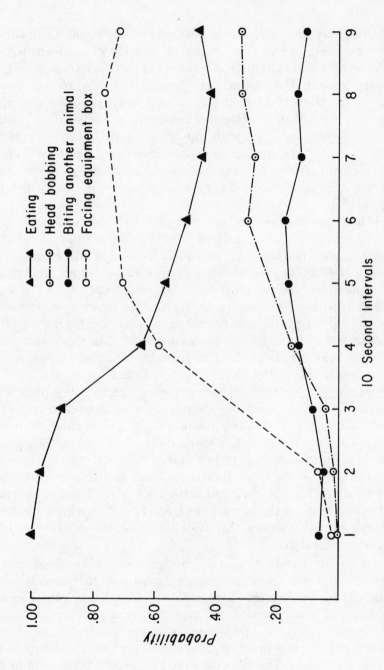

Figure 7.10. Behavior of a female camel (subject 2) during seven sessions on a fixed-time (FT) ninety-second schedule. Also see figure 7.11.

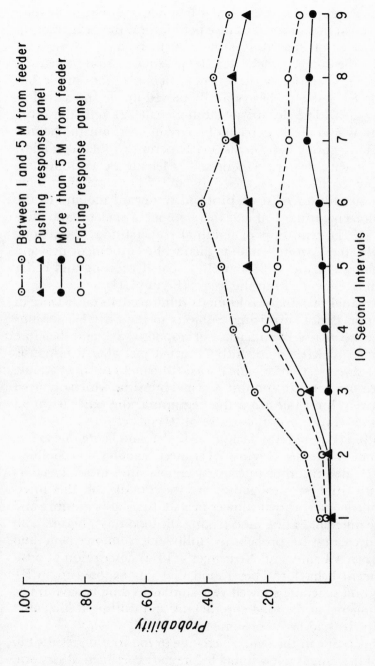

Figure 7.11. Behavior of a female camel (subject 2) during seven sessions on a fixed-time ninety-second schedule. Also see figure 7.10.

227

them food. Baseline data indicated a low rate of biting when food was not present (about one instance per hour for the entire group).

Institution of spaced feeding was marked by substantial changes in the subjects' behavior. In baseline sessions when the feeder was not operational, they spent most of the time outside, but during the FT sessions they typically moved inside and stayed in the feeder area. The dominant adult (female 2) kept the other animals away and ate 96 percent of the reinforcers during the first seven sessions, though she occasionally permitted the other adult (female 3) to eat some of the pellets. The female 2 data from these sessions are plotted in Figures 7.10 and 7.11. Note that some behaviors steadily increase in probability toward the end of the inter-reinforcement interval and therefore meet Staddon and Simmelhag's (1971) definition of terminal (superstitious) behaviors. Specifically, these would include facing the equipment box and head bobbing (Figure 7.10) as well as panel pressing and being more than five meters from the feeder (Figure 7.11).

Since panel pressing was formerly reinforced, its persistence in this situation is not surprising. Subjects in the early FT sessions showed an extinctionlike increase in response rate, and therefore reinforcement delivery frequently occurred just after a response. Even after seven sessions female 2 was still panel pressing at a rate of fourteen per hour. In contrast to panel pressing, which occurred at a relatively high rate from the beginning, the other terminal behaviors developed over the course of several sessions.

Fighting (Figure 7.10), facing the feeder, and being one to five meters from the feeder (Figure 7.11) meet Staddon and Simmelhag's (1971) definition of interim behaviors since these behaviors occur with highest probability in the middle of the inter-reinforcement interval, with lower probabilities at beginning and end of the interval. Eating (and licking the food tray) (Figure 7.10) steadily decreases in probability following reinforcement and therefore meets Penny and Neuringer's (1974) description of post-reinforcement behavior. They found that rats exposed to an FT ninety-second schedule showed very similar feeding behavior; i.e., the rats lingered at the food cup, licking and sniffing it long after the food pellet had been consumed.

At this point in the study, a change in zoo routine resulted in feeding of the camels a few hours before the experimental sessions.

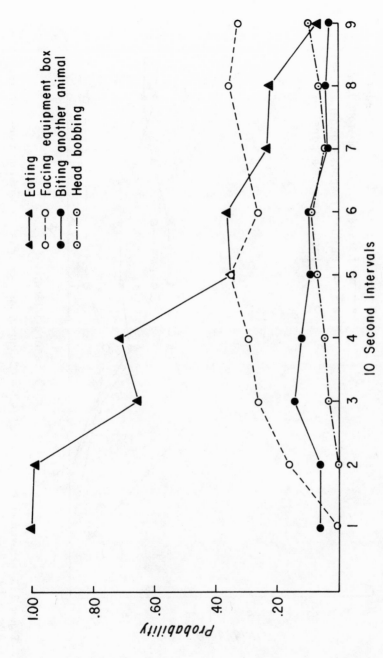

Figure 7.12. Behavior of a female camel (subject 3) during four sessions on a fixed-time ninety-second schedule. Also see figure 7.13.

229

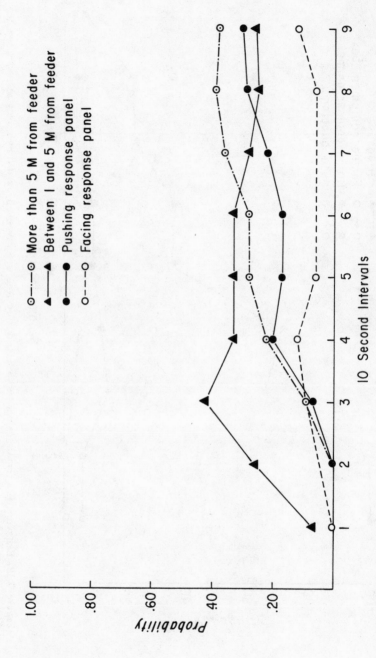

Figure 7.13. Behavior of a female camel (subject 3) during four sessions on a fixed-time ninety-second schedule. Also see figure 7.12.

Legend:

⊙——⊙ More than 5 M from feeder
▲——▲ Between I and 5 M from feeder
●——● Pushing response panel
○---○ Facing response panel

Probability

10 Second Intervals

230

This prefeeding resulted in an abrupt change in the subjects' behavior. Apparently female 2 became satiated to the special food pellets because in the next four sessions she ate only 30 percent of the reinforcers, with female 3 eating 55 percent and the juveniles getting the rest. The female 3 data from these four sessions are plotted in figures 7.12 and 7.13. Terminal behaviors for female 3 included facing the equipment box (Figure 7.13), panel pressing, and being more than five meters from the feeder (Figure 7.13). Fighting (Figure 7.12) and being one to five meters from the feeder (Figure 7.13) are clearly interim behaviors. Eating and licking the food tray again emerge as postreinforcement behaviors.

Head bobbing and facing the key (Figure 7.12) seem to contain elements of both categories (interim and terminal behaviors), perhaps due to the occasional behavior of female 2 in forcing female 3 away from the feeder toward the end of an interval, eating the reinforcer, then moving away.

The slowly decreasing probability of eating following food delivery was seen in both animals and is identical to "postpellet behavior" seen in rats on FT 90 schedules (Penney and Neuringer 1974). After consuming the food pellets, rats typically stay near the food cup, licking it and sniffing at the delivery tube. Very similar behavior was seen in the camels, but has not been reported for birds (Staddon and Simmelhag 1971; Killeen 1975). This difference is probably an artifact of the different food delivery systems used. For rats and camels, food pellets are dropped into a tray and may be eaten at the subject's leisure, whereas pigeons are given brief access to a grain hopper which, at termination of reinforcement, is moved beyond reach.

Another possible effect of these different reinforcement delivery procedures is the temporal patterning of fighting seen in the camels. Killeen (1975) found that pigeons had the highest probability of fighting immediately following reinforcement termination, whereas our camels showed the highest probability toward the middle of the interval. This difference may be due to the camels' tendency to linger at the food tray. Often, their fighting would directly follow such a prolonged eating sequence.

To summarize the results, behavior of the subjects was surprisingly similar considering the different conditions experienced by each individual including changing levels of food deprivation, different hierarchy positions, and uneven previous experience. Although minor differences existed between a few behavior classes, these were no greater than individual differences reported in other studies (Staddon and Simmelhag 1971; Penney and Nueringer 1974). In fact, the behavior of each subject closely resembled the behavior of other species in much more carefully controlled laboratory situations.

GIRAFFES ON FIXED-RATIO SCHEDULES

After our research with camels was progessing well, we decided to expand it to include giraffes (*Giraffa camelopardalis*).[9] We could find no mention of previous operant work with these remarkable creatures, although visual discrimination studies have been reported (Backhaus 1959a, 1959b).

To select an appropriate response, with our camel experience in mind, we carefully watched our giraffes feeding in their barn, begging from zoo visitors, and browsing on trees growing within reach of their enclosure. These observations indicated that giraffes are very sensitive to any movement near their heads and that they usually feed with their lips or tongues. This information, coupled with observations of wild giraffes (MacClintock and Mochi 1973), led us to select a licking response. Based on our experience with the camels, we decided to use contact detectors for manipulanda.

Universal ungulate pellets served as reinforcers, as with the camels. A motor-driven feeder was designed which was fairly quiet in operation. We found a pellet quantity of about 25 grams to be an adequate reinforcer when tested in the subjects' usual feeding area in the barn.

The feeder was installed in a box approximately one meter square and was mounted on a pole about four meters above ground at the edge of the giraffe enclosure. Stainless steel response plates twenty by ten centimeters were mounted on larger Plexiglas panels and placed at the end of pipes about one meter from either side of the feeder. Responses were automatically recorded with conventional relay equipment and, as with the camels, the equipment operation provided auditory feedback for each adequate response.

At the beginning of the study, three giraffes were present in the

9. I thank Georgianne Schmuckal for her assistance in carrying out this research and Tom Cox for his assistance in building the apparatus.

enclosure: Polka, an adult male; Dottie, an adult female; and Ann, their two-year-old daughter. Also in the same African grasslands exhibit were two Kori bustards, four ostriches, and several peacocks. This exhibit was L-shaped with each leg about twenty meters wide, and fifty-five and fifty meters long.

Initial habituation and magazine training took much longer for the giraffes than the camels. Both adults were very cautious in their approaches to the apparatus and would bolt and run away at any movement of the experimenters or loud noise from elsewhere in the zoo. Ann, the juvenile, was much less timid and was finally attracted by carrots.[10] After Ann began reliably approaching the equipment to beg, the adults would come closer and the experimenters could hand them carrots. Ann was the first to be magazine-trained, followed by her mother. Eight daily two-hour sessions were required to shape each of them to lick the panels. Eventually all three giraffes were trained, and each would work for at least short periods; but Polka, the male, was never very consistent. They did not develop a feeding hierarchy and none of them attempted to defend the feeding station. Instead, each would approach only if another animal was not responding.

Ann was the "best" subject and was successfully advanced to an FR 15 schedule (see Figure 7.14 for sample records). An FR 20 schedule of reinforcement produced no stable or consistent responding, and the FR 15 schedule seemed about as much as we could expect from the giraffes because of their relative sensitivity to the distractions so common in zoo research. Ann's response rate increased steadily with larger FR values. She made 0.9 responses per minute on FR 1, 2.1 on FR 2, 4.9 on FR 5, 9.1 on FR 10, 14.2 on FR 15, and 11.4 on a second exposure to FR 10. In contrast, the post-reinforcement pauses (PRP's) tended to decrease as the ratio values were increased. The mean PRP was 65 seconds for FR 1, 58 seconds for FR 2, 49 seconds for FR 5, 44 seconds for FR 10, 44 seconds for FR 15, and 44 seconds for another exposure to FR 10. The response rate data, therefore, were consistent with the expected result, but the PRP data were not. Probably this disparity is a function of the general timidity of these subjects, with the decreasing PRPs

---

10. Carrots are a highly preferred food for most ungulates and would, no doubt, serve as a very effective reinforcer. If we could have developed a convenient way to dispense them, we would have used carrots for such a purpose in this research.

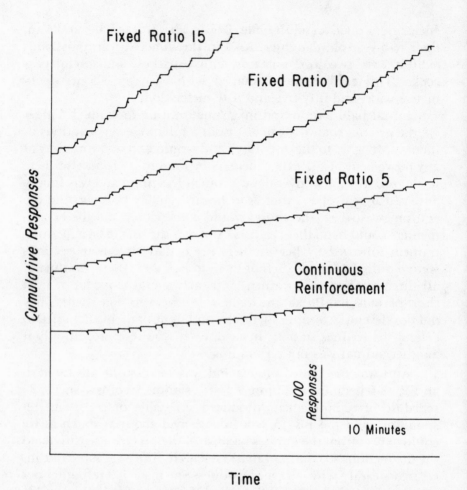

Figure 7.14 Cumulative response records of the stability performance of a juvenile female giraffe on a fixed-ratio reinforcement schedule.

representing a gradual habituation to the experimental setting. This hypothesis is supported by the fact that PRP values did not change in the latter portion of the study when the schedule was changed from FR 10 to FR 15 and then back to FR 10.

The experiment was finally terminated in November (four months after its beginning) because the giraffes would not emerge from their heated barn to work for pellets in the cold Oregon rain. New equipment has been designed for installation inside the giraffe barn, and this research will be resumed in the near future.

OPERANT CONDITIONING OF OSTRICHES

While working with giraffes, we became very aware of the ostriches (*Struthio camelus*) which occupied the same enclosure. Whereas the giraffes were rather shy, the ostriches were quite bold and would approach new equipment without hesitation. Review of the literature revealed no previous work with these giant birds and since they were available, I decided to study them at the first opportunity. The following research was carried out when Wilfried Zimmerman decided to study ostriches for a thesis project (Zimmerman 1974).

We followed the traditional operant approach to birds and selected a heavy-duty, 3.2-centimeter plastic primate key for a manipulandum. The key, mounted on a 1 by 0.5 meter box which was hung from the enclosure fence, was placed about 2 meters high. A small stimulus light was mounted about 30 centimeters above the response key. Informal tests comparing various grains, commerical food pellets, and so on, indicated that the most highly preferred food in the normal zoo diet was raw peanuts. A conveyor-type feeder was constructed to dispense a small handful of nuts on each operation.

In winter 1973 and spring 1974, when this research was conducted, the Portland Zoo collection included one adult male, one adult female, and two juvenile ostriches of undetermined sex. Since they could not be separated, they were studied as a group in daily two-hour sessions. Throughout the study, the stimulus light and response key were illuminated only during these sessions.

After only two days of magazine training, all of the birds, even from as far away as ten meters, would rapidly approach the feeder on each operation. Key-peck training required only minimal shaping, probably because the small size and high visual contrast of the key attracted the birds' attention. For example, even as the equipment was being installed all of the ostriches approached, and several of them pecked the key even before the box had been firmly mounted.

After magazine training was completed, it became clear that the ostriches would not establish a feeding hierarchy as had the camels. Instead, the birds crowded together in front of the equipment, pecking the box and each other and grabbing the dispensed peanuts. The food cup was accessible to several animals at once, and in the confusion some of the peanuts were usually knocked to the ground where they could also be eaten. We were afraid these factors would make shaping difficult, but we encountered no special problems, and after six two-hour sessions, the subjects were responding steadily. Since accurate monitoring of each individual's responding would have been very difficult, we simply recorded the behavior of the group throughout the rest of the study. During shaping, the juveniles were first to start key pecking consistently, and they always showed higher response rates than did the adults; after shaping, the juveniles did most of the key pecking with the adults responding infrequently. Individual response rates in the various conditions were not continuously recorded, but samples of the number of responses made by each individual were hand-recorded at regular intervals throughout the study. Of a total 2,800 responses scored in this way, 44 percent were made by juvenile A, 29 percent by juvenile B, 15 percent by the adult male, and 12 percent by the adult female. Usually the adults would not respond unless the young birds were away from the immediate area.

After the rate of key pecking stabilized, the birds were exposed to a series of fixed-ratio schedules. First, an FR 5 schedule was put into effect for five sessions, then an FR 10 schedule for five sessions; then followed six sessions of FR 15, six sessions of FR 20, and finally five sessions of FR 25.

Cumulative records of the birds responding on FR schedules (Figures 7.15 and 7.16) show the typical post-reinforcement pause followed by a rapid burst of responding for the next ratio (Ferster and Skinner 1957). In this case, however, the PRPs and response rates were much more erratic than those seen in laboratory studies and in the camel data. Also, the usual tendency for total response rates to increase with larger FR values was not evident, with the mean total rate about the same for FR 15 and FR 20 and slightly lower for FR 25. Median total session response rates were eight per minute for FR 5, nine per minute for FR 10, twelve per minute for FR 15, twelve per minute for FR 20, and ten per minute for FR 25.

After the ratio schedules, the birds were exposed to a series of fixed-interval (FI) schedules in which the first response following a specified time period was reinforced. They were first given three sessions of FI 30 (fixed interval of thirty seconds), then ten sessions

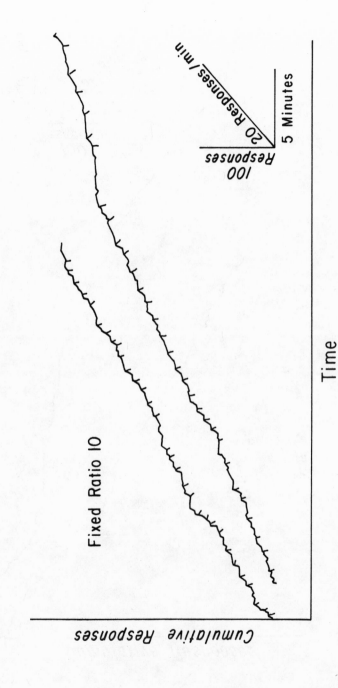

Figure 7.15 Cumulative response records of a group of four ostriches on a fixed ratio 10 schedule. (From Zimmerman 1974.)

237

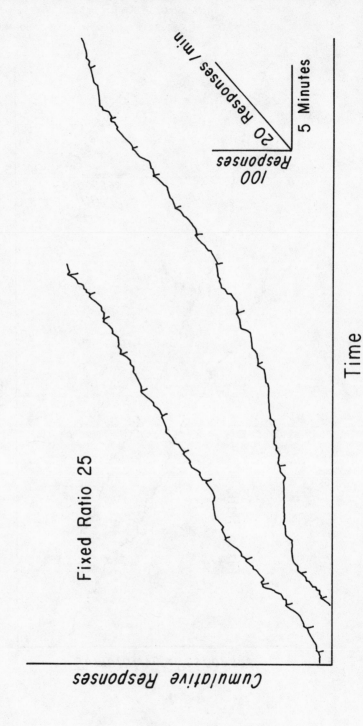

Figure 7.16 Cumulative response records of a group of four ostriches on a fixed ratio 10 schedule. (From Zimmerman 1974.)

238

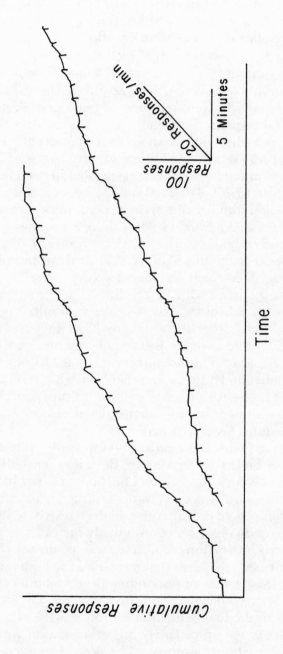

Figure 7.17 Cumulative response records of a group of four ostriches on a fixed interval sixty-second schedule. (From Zimmerman 1974.)

239

of FI 60, and finally seven sessions of FI 90. Next, the group was put on extinction for eight sessions, during which no reinforcers were delivered but responses were recorded.

In laboratory research, the typical response pattern after lengthy exposure to FI schedules consists of a long postreinforcement pause followed by a positively accelerated period of responding, terminating with reinforcement (Ferster and Skinner 1957). Also, the total response rate tends to decrease with increasing FI values. The ostriches did not show the response pattern typically associated with FI schedules. The cumulative records (Figure 7.17) are not much different than those generated under FR schedules (Figures 7.15 and 7.16). Total response rate does decrease, however, with the median number of responses per minute, being 14 for FI 30, 9 for FI 60, and 8.5 for FI 90. Although response rates for FI schedules tend to be lower than rates generated by "comparable" FR schedules (Ferster and Skinner 1957), this relationship is not clear in these data. Such an anomaly could possibly be due to insufficient experience with FI schedules or to the subjects' initial exposure to FR schedules, but it probably results from using a group of subjects rather than an individual subject. In a group situation with other animals also responding and snatching reinforcers, the actual response-reinforcer relationship experienced by any single individual differs somewhat from what a solitary subject would experience with the same equipment program. This factor may well account for the minimal differences between ratio and interval schedule effects seen here.

As observed with the giraffes and especially with camels, the occurrence of breaks in responding decreased markedly over the course of the FR and FI sessions. This finding is not likely caused by a lower rate of distractions in the testing situation because, as the weather improved during course of the research, zoo attendance and its accompanying distractions greatly increased.

When the extinction condition was instituted (by disconnecting the feeder), the birds first responded at a high rate. Later in the session, these bursts of responding became shorter and slower, and pauses between bursts became much longer. The total number of responses in the first extinction session was 155 as compared to 500 responses in the immediately preceding session with an FI 90 schedule (i.e., 1-hour sessions). The second extinction session produced 53 responses; there were 19 responses in the third, 95 in the fourth, and 135 in the fifth. By the sixth session, there were no responses. A rise to 10 followed in the seventh session and, in the eighth, fell back to zero. This is a somewhat lower rate of extinction than is typically shown by individual subjects following FI schedules (Ferster and Skinner 1957); and again, the result may well relate to the complex schedule actually experienced by individuals.

In the last portion of the study, the ostriches were given noncontingent reinforcement. Reinforcers were administered after a fixed interval of time regardless of the subjects' behavior. There were three sessions of FT 30 (i.e., fixed time of thirty seconds) and eight of FT 60. When reinforcement was reintroduced in this form, the animals started key pecking again, even though their behavior did not affect feeder operations. Responding was characterized by long post-reinforcement pauses terminated by short sequences of responding (Figure 7.18). The median total session response rate was 4.5 per minute for FT 30 and 4 per minute for FT 60. Highest rate seen during the FT schedules was recorded in the first session, with the rate slowing and stabilizing in the second session. The birds were still responding at a steady rate of 25 per minute when

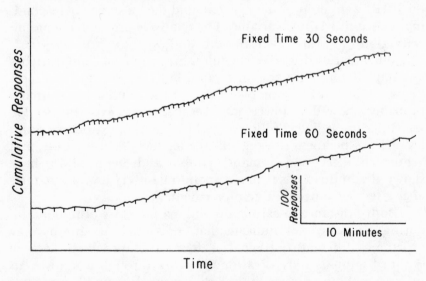

Figure 7.18 Cumulative response records of a group of four ostriches on fixed-time noncontingent reinforcement schedules. (From Zimmerman 1974.)

the experiment was terminated after eleven sessions of FT schedules.

Although systematic observations of individual birds during FT schedules were not made, some clear instances of superstitious behavior did occur. Juvenile B consistently pecked the stimulus light between response bursts on the key, and the adult female frequently turned circles between responses. This latter behavior was probably developed when the bird would stop responding, turn around, and start walking away. When the feeder would operate, she would turn again, run back to the feeder, and usually get a few nuts. The interval was close to timing out by the time she left the feeder again, and she would be reinforced again for turning around.

In general, the data reported here are strikingly similar to those obtained by Grott and Neuringer (1974), who used groups of three rats in a slightly modified Skinner box. In both studies, the post-reinforcement pauses on FR schedules tended to be short and irregular with the total response rate higher than on FI schedules. The rats tended to show steadier and faster responding during FR schedules than the ostriches. Both rats and ostriches failed to develop stable response patterns during FT schedules. In both cases the PRPs were short or nonexistent, and the local rates of responding tended to be quite variable. The relatively greater resistance to extinction was also evident in both studies.

When exposed to FT schedules, groups of rats and ostriches maintained consistent response rates, but the rats showed almost the same rate on FT as on comparable FI schedules. The ostriches probably showed a lower rate because their exposure to FT schedules immediately followed eight sessions of extinction. Studies of FT schedules with pigeons (Zeiler 1968; Lachter, Cole, and Schoenfeld 1971) and humans (Jambor and Stevens 1975) have shown that individuals reduce responding to very low or zero rates after a few sessions of FT reinforcement.

Both resistance to extinction and maintenance of responding during FT schedules indicate that individuals in groups may experience different contingencies than a solitary subject. Because grouped animals only occasionally work through a ratio or an interval without another subject also responding and because not every responding subject in groups gains access to reinforcement on each delivery, individuals are probably experiencing a schedule somewhere between a variable interval (VI) and a variable ratio (VR). Responding after these schedules, especially VI, is very resistant to extinction, and a variable schedule of noncontingent reinforcement quite effectively maintains superstitious responding (Herrnstein 1966; Neuringer 1970; Schwartz 1973).

Comparisons between the findings of Grott and Neuringer's (1974) laboratory study and the zoo-based ostrich research shows the power of simple reinforcement schedules to affect similarly the behavior of groups of subjects, even when dissimilar subjects in substantially different settings are used. Among major variables which do not seem critical for determining group operant behavior are size and shape of the setting, noise level, other organisms in close proximity, quality of the reinforcer, and species of the subjects. In addition, Grott and Neuringer's contention that group performance does not differ in many ways from that of individuals is supported by the ostrich data.

As often happens in research, the results of unexpected events proved as interesting as the main portion of the study. Staff rotations and other disruptions gave us little control over the birds' daily feedings. Occasionally they were fed just before the beginning of a session, and peanuts were not always omitted from the daily ration. At first we were very concerned about this matter, but as it turned out, these feedings had only slight effect on the subjects' behavior. An extreme example occurred several times during the FI and FT schedules when the birds were fed *during* a session. Typically, the ostriches would approach their food trough, eat some peanuts, then run back to the operant station and respond. After eating the earned reinforcers, they would return to the food trough and wait out the interval while sampling their lunch. Because several food cribs were used in the normal feeding proce-dure, the preference for response-contingent peanuts cannot be explained by competition between subjects for the free food. The peanuts dispensed as reinforcers were identical to those given in the regular feedings. Preference for earned over free food has been demonstrated in rats and pigeons under controlled conditions (Neuringer 1969) but has not been reported in regular observations made outside the laboratory. It should be pointed out that prefer-ence for food begged from zoo visitors over the regular diet is a constant problem in the day-to-day care of many zoo animals. Our observations indicate that the type of food thrown by visitors is not the only factor contributing to this preference.

The Zoo Compared to the Natural Environment

To summarize, the results of our experiments were very similar to those of laboratory studies using special strains of rats and pigeons. Although it's unlikely that camels and ostriches will replace rats and pigeons as research animals, the data suggest them as good choices. Further, the zoo experiments demonstrate the considerable generality of behavioral principles developed through laboratory research and the flexibility of operant research techniques. Apparently, at least for some species, many of the control procedures traditionally utilized in the laboratory are unnecessary for obtaining stable, consistent behavior from research subjects.

Taking the issue of generality to the next step, however, raises questions regarding the wild setting. Exhibit enclosures, although much more stimulating than experimental chambers, are still quite barren compared to the natural environment. In the wild there are frequent startling events and many more opportunities to feed, play, explore, or watch other animals than are typically available in the zoo, where distractions so apparent to experimenters are often quite repetitive and probably become "boring" to residents. It can, in fact, be argued with merit that some zoo animals experience substantial sensory deprivation compared to their relatives in the natural habitat. This condition probably increases the reinforcing nature of experimentally controlled contingencies by allowing zoo animals to control at least part of their environment and receive some sensory feedback from operating the equipment (see also chapter 4). This factor may also explain the responding sometimes seen with camels and ostriches in the presence of free food. Still, though far less restrictive than the laboratory, the zoo setting is far from equivalent to the natural habitat.

Surprisingly, very few experimental studies of learning involving free-living wild animals have been conducted; and of these, only a small proportion have been concerned with the basic process of behavioral maintenance and change. Most have been designed to study mechanisms such as internal clocks, navigational systems, social structures, or sensory capacities. For example, in contrast to the volume of laboratory work with rats, the only study of instrumental learning processes with free-living subjects I could find was Ewer's research (1971) on black rats (*Rattus rattus*). She exposed a natural group to various food acquisition tasks that included retrieving peanuts suspended on a piece of string, pulling nuts from a puzzle box, and removing food from a water dish. For each situation, the rats quickly acquired an appropriate response with subsequent gradual change toward more efficient behavior.

Van Lawick-Goodall (1968) reported very similar results for wild chimpanzees in opening special puzzle boxes to gain access to bananas. She suggested that the chimps enjoyed working the puzzle almost as much as eating the bananas. This finding, often reported in studies of free-living animals, indicates that zoo animals may not be very different from noncaptive animals in their willingness to be good research subjects.

Unfortunately, most learning studies conducted in the wild (including the examples cited) have used idiosyncratic techniques which make direct comparisons with laboratory studies difficult or impossible. One notable exception, which illustrates particularly well the potential of working in the natural environment, is Baum's research (1974) using a wild, free-ranging flock of pigeons (*Columbia liva*) as subjects. He installed a standard response panel and feeder in the birds' loft and, using common laboratory techniques, trained them to key peck. Next he exposed them to two response keys, each of which independently controlled reinforcement delivery. Different variable interval schedules were programmed for each key, and the reinforcer frequencies were systematically varied. After a brief habituation period, the pigeons matched their relative response rates to the relative reinforcement rates associated with each key. This result was quite consistent with the "matching law" developed from extensive research in the carefully controlled environment of the Skinner box.

CONCLUSION

The great similarities between results obtained in field, zoo, and laboratory settings show that animals quickly learn to respond appropriately to critical contingencies regardless of considerable environmental differences. This finding supports the assertion that laboratory-derived behavioral principles can be successfully generalized to a wide variety of situations. For example, the effects of reinforcement schedules and the ability to select the richest source of reinforcement are clearly seen in all three settings. Even seemingly minor aspects of behavior such as postreinforcement pauses are usually similarly affected in very different settings. This is to be expected, of course, since natural selection would favor those traits which allowed an organism to habituate to extraneous stimuli while responding adaptively to contingencies controlling food and other survival essentials. Experimenters can control the behavior of their subjects to the extent that they can control these significant events.

Despite the problems listed earlier, zoological parks offer many attractive features to behavioral researchers including accessibility to exotic species and sufficient control of their environment to enable use of most laboratory-developed designs. In addition, since zoos are roughly intermediate between the laboratory and natural environment in terms of environmental complexity, they are particularly convenient places to try out new methods and test the generality of findings from other settings. These factors will undoubtedly attract many more behavioral scientists to zoos in the future.

8 *Vocal Communication in Pinnipeds*

*Ronald J. Schusterman**

INTRODUCTION

The "earless" or "true" seals, the sea lions, fur seals, and the walruses make up the three families within the order Pinnipedia (see King 1964 and Scheffer 1958 for detailed information on taxonomy and zoogeography). Like the whales, porpoises, and dolphins (order Cetacea), the pinnipeds are large-brained marine mammals which have long held paradoxical fascination for man— paradoxical because their appeal is aesthetic as well as commercially exploitive. Recently, several pinniped species have been subject to rather intense scientific study in the field and captivity. Intensified study of these amphibious mammals of the sea was undertaken for several reasons, not the least of which include their relatively long history (about 200 years) of commercial exploitation and their little-known role in marine ecology. A fruitful exchange of information and ideas about pinnipeds recently occurred at a symposium on "Biology of the Seal" at the University of Guelph, Ontario, Canada, in August 1972. Researchers gathered here from North and South America, Europe, the Soviet Union, Japan, and Australia to discuss the evolution, zoogeography, ecology, anatomy, physiology, behavior, and husbandry of seals, sea lions, and

*Ronald J. Schusterman is a member of the Departments of Psychology and Biology at California State University, Hayward, California. This research was supported by ONR Contract N00014-72-C-0186.

walruses. Some of the present chapter is based on the stimulating discussions of pinniped behavior (particularly social systems and sensory systems) generated by fundamental research on both natural-living and captive populations. Such information is quite relevant to continuing efforts in conservation and management of the pinnipeds.

Ideally, pinnipeds, like other animals, are best observed and experimented with in their natural habitat; but detailed observations plus control and manipulation of key variables can often be accomplished only in a captive situation (Schusterman 1968). As pointed out by Eisenberg (1966), laboratory and field studies of wild animal behavior should be integrated so that problems identified in the field can be isolated and experimentally investigated in the laboratory; and, conversely, so that laboratory findings may be verified in the field.

Contemporary species of pinnipeds feed mostly on cephalopods, crustaceans, mollusks or shellfish, and fish. They combine two characteristics which occur together in no other mammalian species: they reproduce (sometimes copulate and always nurse their young) on land or ice; and they are third-level consumers in the marine food chain.

Although seals and sea lions have performed and been on display in zoos, aquariums, and circuses all over the world for over one hundred years, only since the early 1960s have the vocal communication patterns, learning, and sensory perception of captive animals been studied to an appreciable extent. A few investigators located in Canada, Denmark, and the United States have accomplished the bulk of this research. My own research with captive pinnipeds has concentrated on the behavioral and physiological mechanisms by which information is acquired and transmitted for their succesful adaptation to both terrestrial and marine environments (Schusterman 1968, 1972). This research made it clear that visual and acoustic communication were important in both the atmosphere and the hydrosphere. The purpose of this chapter is to stress significance of the vocal-auditory communication system in several pinniped species. In this task, it is necessary to identify the selective pressures rising from the particular nature of their habitat (e.g., surf noise) which may have been critical in the evolution of the design features of the communication system.

VOCAL DISPLAYS IN THE WILD AND IN CAPTIVITY

It is not my intention to review the vocal repertoires of the pinnipeds (see Winn and Schneider in press). Suffice to say that in all polygynous species, where females clump together and males compete for access to females, both males and females as well as pups are quite vocal. As far as is known, the sea lions and fur seals show relatively extreme forms of polygynous breeding structure. Moreover, at least three species of the "earless" seals also maintain an organized polygynous mating system in which females clump together. In all of these species there is terrestrial parturition, nursing, and copulation, and a problem regarding habitat availability and crowding. It seems probable that cómpetition for space has played an important role in the evolution of vocal signalling as a dispersal mechanism in males. Moreover, in some species male calling may also play a significant role in the female's choice of a mate.

With the constant threat of mother-pup separation, crowding has probably acted as a selective pressure on the parturient females toward frequent calling to their newborns and toward rapid learning to recognize calls of their own pups (Petrinovich 1974). Generally then, either males compete for the right to copulate with gregarious females by dividing the breeding grounds through a quite rigid or flexible territorial behavior; or they compete for females by maintaining individual distances contingent on a pecking order or social hierarchy (Bartholomew 1970). In either case, the social systems are maintained primarily by vocal signalling (although postures, facial expressions, and movements are also important in threat and submission) reinforced by physical aggression (pushing and biting). Generally, distinct sexual differences exist in the vocalizations of polygynous breeding pinnipeds. In some cases, like that of the California sea lion, the difference is quantitative rather than qualitative, with males barking much more frequently than females.

Importance of the vocal-auditory channel in some pinniped forms can be readily appreciated by even the most casual observer of captive or wild animals. Exploring the northern California coast during the winter and spring, one can hear incessant raucous barking offshore by male California sea lions (*Zalophus californianus*) or the honking "clap-threat" of male northern elephant seals

(*Mirounga angustirostris*). These sounds are both loud and directional, at least to the human ear.

One major advantage of the aerial vocal-auditory system of communication in pinnipeds is that messages can be conveyed over a relatively long distance, in the darkness or fog, and around obstacles, thus minimizing the large energy expenditures necessary for the awkward locomotion of pinnipeds on land. Peterson and Bartholomew (1969) point out that many pinnipeds have anatomical adaptations that enhance sound production and that selective pressure imposed by the high level of ambient noise in pinniped rookeries and hauling grounds is largely responsible for the frequent repetition, unusual loudness, and apparent conspicuous directionality of their vocalizations.

In order to communicate the flavor of airborne vocalizations in pinnipeds so that you can appreciate their significance, the following brief descriptions are quoted from both field and captive observations.

> California sea lion (*Zalophus californianus c.*):
> There may be no other mammals that vocalize as loudly and incessantly as territorial male California sea lions. Their vocalizations consist of a series of *barks*, which at the height of the breeding season are repeated almost continuously, day and night. . . . During the breeding season barks appear to serve primarily as threats. They occur as preludes to ritualized displays and to overt fights. However, bulls bark vigorously when approaching females. . . . A bull with an aquatic territory may lift its head above the surface and bark a few times, then resubmerge. . . . Regardless of its posture, a barking bull usually holds it mouth half open [and] extends its vibrissae forward. (Peterson and Bartholomew 1969. pp. 17–18.)

> Steller sea lion (*Eumetopias jubatus*):
> While on his territory a male prefaced every movement with a vocalization. An animal lying prone would vocalize with a roar when he rolled onto his side and would roar again when he stood up. The exaggerated walk was accompanied by a prolonged belchlike roar. . . . Sometimes the roar or pulsed roar alternated with a hissing exhalation of air which was audible up to a hundred meters away. (Gentry 1970, p. 24.)

> Elephant seal (*Mirounga angustirostris*):
> The most characteristic vocalization of the male elephant seal is a formalized clap-threat. . . . [It] is used as a signal of incip-

ient attack, and it is a much higher intensity threat than the snort. It is the primary signal for readiness to fight and is always produced from a formalized, stereotyped posture in which forequarters are elevated to the maximum and then inflated . . . proboscis extends down into the fully open mouth. . . . The clap-threat is an extremely loud, resonant, clapping sound with a metallic quality which suggests the exhaust noise made by a diesel engine. (Bartholomew and Collias 1962, p. 11.)

Gray seal (*Halichoerus grypus*):
We recently recorded a male and female when they were six or seven months of age. . . . Continuous high-pitched humming or moaning sounds, which sound something like a dog crying in its sleep, sometimes with a degree of frequency modulation giving it a slight yodeling quality . . . have also been recorded. . . . (Schusterman, Balliet, and St. John 1970, p. 304.)

It is thus clear from both field and laboratory observations that natural selection has favored the vocal-auditory communication channel in pinnipeds. But several questions arise regarding the extent to which this communication channel is operative in submerged pinnipeds. Also, how directional are airborne threat vocalizations and over what spectral range? Do the vocal threat displays of the California sea lion or elephant seal bulls have a constant form or "typical intensity" (Morris 1957), or are these displays graded so as to reflect changes in the animal's motivational state? Do vocalizations change seasonally? To what extent and under what conditions does the vocalizing animal affect the behavior of other animals? How efficient is the auditory reception of sounds? To what extent are pinniped vocalizations available for operant conditioning? Certain aspects of these questions can be answered in the field, but frequently the refined measurement techniques and experimental controls available only in the laboratory are necessary for grappling with such problems. In the next two sections of this chapter, research dealing with these issues will be reviewed, and previously unpublished research on the vocal behavior of captive California sea lions will be presented.

Underwater and Aerial Vocalization

Pinnipeds are now known to produce a wide variety of vocalizations below as well as above the water's surface. Their underwater sounds have been described as loud barks, whinnies, faint clicks, trills, moans or humming sounds, chirps, belches, growls, squeals, roars, roarlike growls, etc. In general, spectrographic analyses reveal that the main acoustical energy components range from about 0.1 to 10 kilohertz. The structure of sounds emitted by seals and sea lions differs from that of porpoises and whales in a number of parameters. Comparison in terms of the spectral characteristics of the acoustical energy reveals, for example, that in contrast to the generally low-frequency clicks emitted by pinnipeds, the clicks of dolphins may contain a broad frequency spectrum ranging up to 256 kilohertz with maximum energies between 20 and 60 kilohertz. Emission of whistles in dolphins is quite different in frequency characteristics from almost any sound produced by the pinniped forms. Porpoise whistles contain a very narrow frequency band and are modulated with a frequency range of approximately 4 to 20 kilohertz. In terms of temporal parameters, a series of clicks by pinnipeds rarely contains an individual pulse duration of less than 0.005 second or a repetition rate of more than 50 to 150 clicks per second, whereas porpoise clicks usually have a duration of 0.001 second and repetition rates of over 500 per second (see Kinne 1975 for a review on sound production in marine mammals).

Despite a plethora of recordings and spectrographic analyses of underwater vocalizations by seals and sea lions in both captivity and the wild, sampling of the underwater vocal repertoire of most species has been quite limited, and the adaptive significance of such calls has remained relatively obscure. Structural and temporal properties of underwater vocalizations, however, are similar to airborne calls (Schusterman and Balliet 1969; Schusterman, Balliet, and St. John 1970).

The function of airborne calls by pinnipeds has been much easier to study than the function of underwater calls, since in the former case an observer has much better opportunity to see, hear, and identify individual animals and to associate vocalizations with other specific behavioral or environmental events. Therefore, since wild and captive studies have repeatedly demonstrated that airborne vocalization by seals and sea lions functions either to threaten or attract individuals by identifying the caller as to species, sex age, location in space, and probably as an individual (Bartholomew and Collias 1962; Peterson 1968; Schusterman and Balliet 1969; Stirling and Warneke 1971; Petrinovich 1974; Sandegren in press), it appears likely that underwater calls serve the same socially communicative functions. This statement may be contrasted to the notion that pinniped underwater sound production principally functions for echolocation (Poulter 1963).

Two brief examples (one each from the field and the laboratory) serve to illustrate this outlined "amphibious" approach to the structure and function of pinniped vocalizations.

Underwater sound emissions by Weddell seals (*Leptonychotes weddelli*) were first described structurally by Schevill and Watkins (1965). But it was not until Ray (1967), Kooyman (1968), and Kaufmann, Siniff, and Reichle (1972) observed that underwater postural displays and fighting as well as some copulatory behavior were associated with "chirping" and other vocalizations (such as the "trill" and "chugs") that these underwater calls were clearly understood to function as threat and possibly sexual displays, particularly by adult males. All such vocalizations and associated postures and movements by the Weddell seal have been observed to occur *on* as well as *under* the Antarctic ice.

Schusterman (1968) showed that in captivity removal and reintroduction of alpha male California sea lions serve to disinhibit or inhibit barking in the beta male. Experimental manipulations clearly demonstrated the significance of loud barking in the resident male's threat pattern. In the same experiment, barking by the seal lions frequently occurred while their heads were submerged. It was concluded that barking functioned underwater as in air, i.e., as threat displays by neighboring territorial males which decreased overt physical aggression (Schusterman and Balliet 1969).

DIRECTIONALITY

In the polygynous species of pinnipeds, male threat vocalizations are usually very loud and repetitive, have a rapid onset, and often have an effective frequency range from approximately 0.1 to 10 kilohertz. The most characteristic threat vocalization by the male northern elephant seal, for example, is the so-called "clap-threat." Acoustically, the calls are highly pulsed with each pulse energy concentrated below 2.5 kilohertz. Bartholomew and Collias (1962) describe how one bull elephant seal *directs* its clap-threat at another. Observations in the field by Peterson and Bartholomew (1969) indicate that the bark of the California sea lion is an extremely directional vocalization—so directional, in fact, that Peterson and Bartholomew, when groping around the rookery at night, could not only locate a barking bull by acoustical cues alone, but could also locate the direction of its movement and the specific area of territory on which it was focusing its barks.

The bark of *Zalophus* bulls is characterized by a rich harmonic structure and an up-down-up frequency modulation. Most acoustical energy is below 4 kilohertz with relatively little variation in the duration of a single bark which lasts 200 to 300 microseconds (see Figure 8.1). Sound localization depends mostly on differences in the sound input to the two ears. Other sea lions can probably readily localize a series of underwater or airborne barks by binaural detection of differences in intensity and time of arrival (Moore 1975; Moore and Au 1975).

During the breeding season, one can usually stimulate a captive *Zalophus* bull to focus his barks directly at one merely by standing in front of him. The sound is positively deafening even at a distance of only a few meters. Like Peterson and Bartholomew (1969), I was very impressed by the degree to which the bark of *Zalophus* is a highly directional phonation. In an attempt to quantify the directionality of this phonation, five captive *Zalophus* were stimulated to bark repetitively from a fixed distance into a microphone which was oriented at three different positions in the median plane. In order to obtain comparative data on the bark phonation, the experiment was repeated with a 6-year-old husky-type dog (Norwegian elkhound). The procedure involved changing the composition of a captive *Zalophus* group so that only a single animal emitted barks (see Schusterman 1968; Schusterman and Dawson 1968). Sound pressure levels (dB re .0002 dynes/cm^2) were measured with a Bruel and Kjaer precision sound level meter (type 2203) in three adjoining outdoor compounds, separated from one another by cyclone fencing; since no obstructions such as brick walls hindered them, sound level measurements were considered to have been made in a relatively free field. The three compounds in

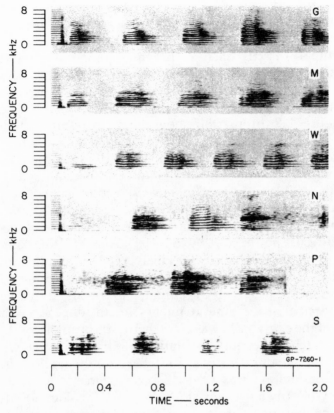

Figure 8.1 Spectrographs of aerial barks by male *Zalophus*. The signals are extended in time because of reverberation.

which measurements were made were approximately 8 by 18 meters, 4.5 by 4 meters and 6 by 8 meters. A condenser microphone (type 4131) was used in conjunction with the sound pressure level meter. The microphone, with a linear frequency response from 20 hertz to 15 kilohertz to within ±1 decibel, had equal sensitivity for sound coming from all directions (i.e., was completely omnidirectional), from 20 hertz to 3-4 kilohertz. Above 4 kilohertz, the change in frequency response as a function of sounds coming from different angles was considerable. An octave analysis was made by using the sound pressure level meter with an octave filter set (type 1613) which contained eleven band-pass filters.

Barking intensity by all individuals was measured at a distance of approximately two meters, and at least twelve readings were taken at each band-pass filter setting in three orientations: (a) with the animal directly facing the microphone (0° orientation); (b) with the animal facing at a right angle to the microphone (90° orientation); and (c) with the microphone placed directly behind the head of the animal (180° orientation).

Figures 8.2 through 8.7 show the main results of the experiment. The average barking intensity is given both as a function of orientation of the sea lion's head to the microphone and as a function of the analyzing filter band. These results may be summarized as follows: (1) The greatest acoustical energy in the *Zalophus* bark is at 1 kilohertz with rapid intensity drop-offs below 500 hertz and above 2 kilohertz. (2) The main energy of their barks when directed at a receiver two meters away at 1 kilohertz ranges between 103 decibels and 95 decibels. (3) The average decline in intensity at 1 kilohertz for all five sea lions from 0° to 90° orientation was 6 decibels as compared to a drop-off in intensity of only 1 decibel at the same frequency for the dog. Since the main energy of the dog's barks was at 500 hertz, the drop-off for the dog is 3 decibels when the same orientation comparison is made. (4) The average decline in intensity at 1 kilohertz for the sea lions from a 0° to 180° orientation was 12 decibels as compared to a decline in intensity of only 5 decibels at 500 hertz for the dog. (5) Although the more mature California sea lions did bark louder than the younger animals, directionality did not appear to be related to sex.

This experiment supports two notions proposed by Peterson and Bartholomew (1969): first, that the conspicuous directionality as well as the unusual loudness and repetitiveness of some pinniped vocalizations (especially threat) "have been favored by selective pressures imposed by the high level of background noise on land." Peterson and Bartholomew also pointed out that in *Zalophus* the arytenoid cartilages have been modified for the production of loud sounds. It appears likely that other structural modifications of their vocal apparatus have enabled *Zalophus* to emit an especially directional airborne bark. Second, the emission of narrow cones of aerial sounds allow *Zalophus* to address the

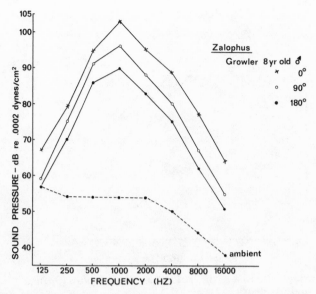

Figure 8.2 Average intensity of the barks of a male *Zalophus* (G) as a function of directionality and frequency.

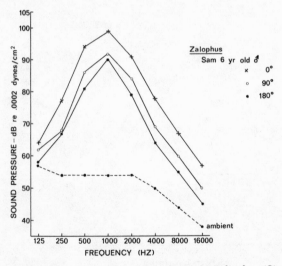

Figure 8.3 Average intensity of the barks of a male *Zalophus* (S) as a function of directionality and frequency.

257

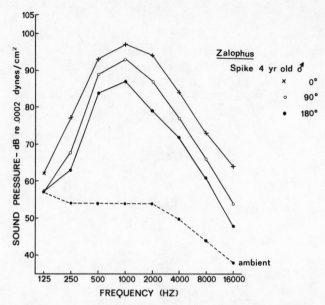

Figure 8.4 Average intensity of the barks of a male *Zalophus* (Sp) as a function of directionality and frequency.

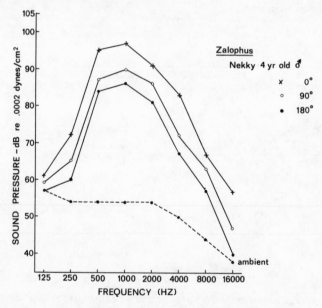

Figure 8.5 Average intensity of the barks of a male *Zalophus* (N) as a function of directionality and frequency.

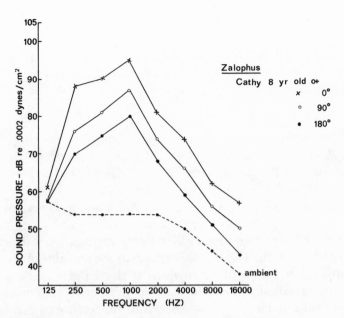

Figure 8.6 Average intensity of the barks of a female *Zalophus* (C) as a function of directionality and frequency.

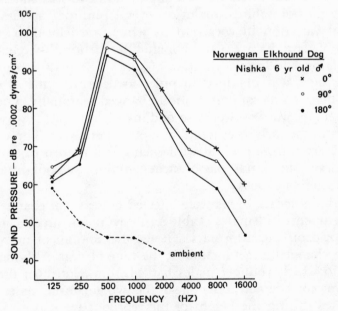

Figure 8.7 Average intensity of the barks of a male Norwegian elkhound dog (N) as a function of directionality and frequency.

259

loudest portion of its signal toward the intended recipient, thereby threatening or attracting specific individuals rather than whole groups of animals.

SEASONAL VOCALIZATION

In the wild, only one species of pinniped, the ringed seal (*Phoca hispida*), has been observed long enough to quantify vocalizations both during and outside the reproductive season (Stirling 1973). The ringed seal emits several vocalizations under the Arctic ice, including yelps and barks, and the latter sound is similar in acoustic structure to the threatening underwater bark of *Zalophus* (Schusterman and Balliet 1969). From winter to spring, Stirling (1973) found 11 percent increase in the amount of vocalizations and related this phenomenon to the ringed seal's greater degree of agonistic behavior during the reproductive season. Seasonal cyclicity has also been shown in the territorial displays of a captive group of *Zalophus* (Schusterman and Gentry 1971).

These observations indicate that, similar to seasonal singing of male passerine birds, males of several pinniped species show seasonal variation in vocalizations which are related to various aspects of reproductive activity including territoriality. Quantitative data on barking vocalization of six captive male *Zalophus* of varying ages was obtained in order to determine the extent to which their vocal communication changed seasonally. Vocalizations were recorded at twenty-second intervals at least twice weekly (usually in the early afternoon) for a period of a year. The animals were observed from a blind while being maintained in two relatively large compounds, one of them containing a pool (for details see Schusterman 1968).

Barking vocalizations were affected by seasonal changes despite the absence of females. Table 8.1 shows that in an undisturbed group of captive California sea lions, the amount of individual barking changed as a function of the time of year. Although the degree to which younger males barked from September through April was considerable, only one of the two territorial males (G or M) barked during the height of the reproductive season, i.e., in June and July. Notice that outside the reproductive season, the older males do most of the barking. Barking is related to territorial behavior during the breeding season; while, during the nonrepro-

Table 8.1 SEASONAL VARIATIONS OF BARKING VOCALIZATIONS
IN A SOCIAL GROUP OF MALE ZALOPHUS

APPROX. AGE (YRS)	G (6-7)	M (6-7)	W (4-5)	N (4-5)	P (4-5)	S (3)
	% Bark	% Bark	% Bark	% Bark	% Bark	% Bark
DATES 1968						
Jun-Jul	22	1	0	1	0	12
Aug-Sept	24	12	15	6	1	7
Oct-Nov	31	44	30	10	5	2
Dec-Jan	42	17	40	9	7	9
1969						
Feb-Mar	32	22	27	17	16	2
Apr-May	16	2	45	24	7	3
Jun	5	45	1	0	0	0

ductive season, barking is related to a weak, size-related dominance hierarchy.

In June and July 1968, G was the dominant male of the group, and the other larger males clearly inhibited barking in his presence. Only the youngest animal in the group, S, could bark in G's presence without being attacked. This period coincides with the reproductive activity of *Zalophus* in the wild. At the end of the reproductive season in August-September, the other males of the group began vocalizing to a much greater extent than previously. This release from vocal inhibition by the other group members continued until May, when subdominant male M began inhibiting in the presence of the dominant male G. During the last week in May, however, M challenged G and acquired the pool, driving G into an area without water. M then became the dominant animal within the pool area, and barking by all other animals was inhibited.

As a supplement to the laboratory work, preliminary observations on vocalization and social behavior of *Zalophus* males during the *nonreproductive* season were made by Roger Gentry at Año Nuevo Island (45 miles south of San Francisco, California). The purpose of this work was to ascertain whether several aspects of the vocal behavior of captive males resembled those of their wild counterparts. During the nonreproductive season, a size-related dominance system existed among the males. Larger animals always prevailed during aggessive encounters and had the prerogative of vocalizing. Although small or submissive males inhibited barking vocalizations, they nevertheless barked to some extent outside the reproductive season just as they do in the laboratory. In addition to barking, two kinds of vocalizations, infrequently heard among captive animals, occurred among the feral males. These sounds, the "roar" and the "whinny," were not as prevalent as barking and were associated with more intense aggression than was barking. Captive females make both of these sounds, both in air and underwater (Schusterman, Gentry, and Schmook 1966). Roars and whinnies occurred during competition for resting sites where large numbers of males were onshore together, but seldom if ever occurred when an animal was alone.

Males of all sizes and ages produced any of the three kinds of vocalization, although they apparently did so with different frequencies. For instance, barking was the predominant vocalization among the largest males, occurring in specific social encounters as well as in nonspecific situations. Episodes of barking seemed less intense and of shorter duration in the latter context. This is in marked contrast to the almost incessant barking by territorial males during the reproductive season.

Barking in *Zalophus* is primarily a male vocalization and shows seasonal variation in relation to territorial behavior similar to the singing of passerine birds. The fact that territorial males bark almost continuously—not only when actively chasing male intruders, patrolling their territorial borders or herding females, but also seemingly "self-advertising" in the center of their territory (see next section)—suggests that barking, like male bird singing, may play a significant role in attracting mates (Kok 1972).

Since female *Zalophus* usually copulate only once (Peterson and Bartholomew 1967) and the males' only role in reproduction is to provide a set of genes (but see Barlow 1972, 1974), females may be making rather fine judgments about a male's relative fitness on the basis of quantity as well as loudness of his barking. To my knowledge, the question of whether the vigor of a territorial male's vocalization is in fact correlated with his reproductive success has not as yet been systematically investigated. However, Cox and LeBoeuf (in press) indicate that the loud threat vocalizations of northern elephant seal females during attempted mounts by all males results in a greater likelihood that mature males of higher social rank will complete their attempted copulations (see LeBoeuf 1974). Thus in the nonterritorial polygynous breeding system of the northern elephant seal, the degree to which a female selects an optimal male genotype depends to some extent on the vigor of her protesting threat vocalization.

INFORMATION CONTENT

Sandegren (in press) has recently elaborated on the original work by Bartholomew and Collias (1962) regarding the three types of threat vocalizations by northern elephant seal bulls and the postures associated with these vocalizations. He notes, for example, that both the probability of occurrence as well as intensity of the "snort," often emitted from a prone position by the dominant male, increases as a subdominant male approaches. Intensity of the "snort" vocalization is thus directly related to the degree to which the adult male *Mirounga* is prepared to fight.

On the other hand, the "clap-threat" (which is the most prominent vocalization) is considered by Sandegren to function primarily for individual recognition and only indirectly as a threat by identifying the individual's previously acquired status. Thus the parameters of this pulsed vocalization, probably important in individual recognition, are structural (frequency, duration, harmonic embellishments, etc.).

According to Sandegren, each bull emits a "signature" vocalization (clap-threat) in connection with fighting and associates a particular vocal signature with the outcome of the fight. Thus, the frequent and intense mutual vocalization prior to an impending fight is believed to diminish the likelihood of a high-ranking male being repeatedly challenged by mistake.

Although this signature vocalization shows much variation between individuals, it is highly stereotyped within an individual, particularly with regard to temporal patterning. Apparently, the temporal patterning of the so-called "clap-threat" vocalization remains constant no matter what the circumstances. It is, therefore, unlikely that this vocalization transmits information regarding the moment-to-moment fighting motivation of the individual. LeBoeuf and Peterson (1969) found that the pulse repetition rate of the vocalization was extremely stereotyped within a given male colony and suggested that such geographical differences in vocal behavior are similar to local dialects in birds and humans. For example, the mean pulse rate from the Año Nuevo Island colony was 1.02 (with a standard deviation of 0.11), and this pulse rate was significantly different from the mean pulse rate of 1.88 (with a standard deviation of 0.41) from the San Miguel Island colony. This pulsed threat vocalization of *Mirounga angustirostris* thus has a constant form or "typical intensity" (Morris 1957) rather than a graded form which would reflect the degree to which the animal was prepared to fight. The primary information conveyed by the "clap-threat" vocalization is apparently related to an individual's social status within the dominance hierarchy. Instead, aside from the "snort" vocalization, the graded aspects of the northern elephant seal's displays are visual, associated with neck and body elevation, head tossing, "slamming" of the body on the ground and, in particular, by the shape of the proboscis (see Figure 8.8). It is through these graded visual displays that the fighting motivation of an individual can be evaluated (Sandegren in press).

Figure 8.8 Elephant seal bull emitting a "clap-threat" vocalization with proboscis fully inflated.

Whereas territorial *Zalophus* bulls primarily use a single kind of vocalization (the bark) during the *reproductive* season, the territorial male northern fur seal (*Callorinus ursinus*) uses four readily distinguishable vocalizations, each in a particular situation (Bartholomew 1953; Peterson 1968). *Callorinus* bulls lunging at each other in a boundary display emit an explosive "puff" vocalization; whereas a clicklike vocalization called the "whicker" is continually emitted when a bull gets up from a resting position and locomotes around his territory (Peterson 1968). In similar fashion, territorial male Australian fur seals have a considerably richer repertoire of threat vocalizations than does *Zalophus*; these include barking, guttural threats, low-intensity threat calls, and whines or appeasement calls (Stirling 1971; Stirling and Warneke 1971).

The question thus arises as to how territorial *Zalophus* bulls are able to grade their barking vocalizations in order to transmit enough information to permit maintenance of breeding territories similar to northern and southern fur seals. Peterson and Bartholomew (1969) have suggested that *Zalophus* bulls grade their vocalization by varying the direction and therefore the loudness (see previous section on Directionality) of their barks, changing both the number of barks in a series and the rhythmic pattern of barking. To support this notion, these researchers analyzed forty-five series of barks and found that when a bull threatened a specific opponent in a vocal display, the mean number of barks per series was 11.1, whereas in routine patrolling of a terrestrial territory it was 8.1. In terms of varying the rhythmic pattern, they believed that rapid barking occurred during "vigorous close-range interactions" between adversaries, whereas a slower rate was manifested "in calmer, more placid situations" when a single bull is routinely self-advertising. Unfortunately, Peterson and Bartholomew presented no quantitative data to support this observation.

In order to obtain quantitative data on whether the temporal patterning of barks is an important carrier of information in the vocal communication of *Zalophus*, both aerial and underwater barking vocalizations were recorded on magnetic tape from a captive group of bulls during the reproductive season. The primary analysis was accomplished with a Raytheon Rayspan spectrum analyzer. Magnetic tape-recorded signals of the barks fed into a scanning tube emerge with a continuous spectrograph at either a real time base or at some fraction of actual recording speed (Poulter 1968). The temporal parameters analyzed included:

(a) Number of barks in a barking series, which is defined as a succession of barks followed by an interval twice as long as all preceding interbark intervals of that series.

(b) Duration of a series of barks with the beginning of the first bark and the end of the last bark.

(c) Changes in the repetition rate of barks.

(d) Single bark duration.

While recording vocalizations on one track of the tape, the observer identified on the other track: the sea lion that was barking, whether it was stationary, locomoting slowly, or galloping; whether the barking was directed at another individual; and whether a "boundary ceremony" was involved (see Figure 32 in Schusterman 1968). Two situations prevailed during these observations. For the 1969 group consisting only of *Zalophus* males (including two adults) and two yearling Steller sea lions as well as harbor seals, only a single pool was available. In 1970, the group consisted of California sea lions only. Two adult males and two juvenile males were housed together with four females, and two pools were available.

Tables 8.2 through 8.4 show the data obtained on one day in 1969 with a group composed only of male *Zalophus*. The two adult males, Growler and Marty, were equally dominant at this time and had set up territories within "compound 1."

Table 8.2 Analysis of California Sea Lion Growler's Barking Pattern in Compound 1 During an 8-minute Observation Period.
(Date: 5/29/69   Time: 15:00)

	No. of Barking Series Directed by G							
	California Sea Lions					Stellers	Nondirected	Total
	M	W	N	P	S			
Locomote	11	2	2	0	0	0	0	15
Stationary	34	0	0	1	0	0	0	35
Total	45	2	2	1	0	0	0	50

	Directed Barks			
	No. Barks/Series		Barks/Sec. (Rep. Rate)	
	$\bar{X}$	SD	$\bar{X}$	SD
Locomote	8.8	3.7	2.4	0.5
Stationary	5.9	1.9	2.3	0.4

Table 8.2 shows a sample of temporal barking patterns (see Figure 8.9) by Growler when barking was not included in the context of a boundary ceremony. These data can be summarized as follows: In this sample, (a) all of Growler's barks were specifically directed at other males, and he never barked in a routinely self-advertising fashion; (b) Growler directed forty-five of fifty, or 90 percent, of his series at the only other adult male ("M" or Marty); (c) Growler's mean number of barks per series was significantly greater when he directed his barks during locomotion than when he was stationary (*P*'s < .05, t-test) regardless of whether a bark series were directed or nondirected; and (d) the mean repetition rate of Growler's barks was the same whether he was stationary or moving.

Table 8.3 ANALYSIS OF CALIFORNIA SEA LION MARTY'S BARKING PATTERNS IN COMPOUND 1 DURING AN 8-MINUTE OBSERVATION PERIOD
(Date: 5/29/69   Time: 15:10)

	NO. OF BARKING SERIES DIRECTED BY M							
	California Sea Lions					Stellers	Nondirected	Total
	G*	W	N	P	S			
Locomote	0	4	5	4	2	0	0	15
Stationary	45	11	9	9	6	0	21	101
Total	45	15	14	13	8	0	21	116

	DIRECTED BARKS			
	No. Barks/Series		Barks/Sec. (Rep. Rate)	
	$\overline{X}$	SD	$\overline{X}$	SD
Locomote	10.5	5.1	2.7	0.3
Stationary	7.4	3.4	2.7	0.3
	NONDIRECTED BARKS			
	No. Barks/Series		Barks/Sec. (Rep. Rate)	
	$\overline{X}$	SD	$\overline{X}$	SD
Stationary	7.1	3.0	2.0	0.3

*M directed barks while in compound 1 at G who was behind fence in compound 2.

SECONDS

Figure 8.9 A sample of two series of barks analyzed by the Raytheon Rayspan spectrum analyzer. Sample *a* shows a series of barks with a repetition rate of 2.7 barks per second, while *b* shows a series of barks with a repetition rate of 1.8 barks per second.

Table 8.3 shows a sample of temporal barking patterns by the other adult male Marty, when barking was not included in the context of a boundary ceremony. Marty's data can be summarized as follows: In this sample, (a) Marty's barks could be classified as both directed and nondirected; (b) Marty directed forty-five of ninety-five, or 47 percent, of his directed bark series at the other adult male ("G" or Growler); (c) like Growler, Marty's mean number of barks per series was significantly greater when he directed his barks in locomotion than when stationary ($P < .05$); and (d) although the mean repetition rate of Marty's barks within a series was the same whether he was stationary or moving, the mean repetition rate within a series classified as nondirected was significantly slower than the mean repetition rate of a series of directed barks (P's $< .01$).

Table 8.4 compares barking patterns when the two adult males were barking at each other at close range near or at the border of their territories or barking from approximately the center of their territories (usually in a stationary position). Although both sea lions showed a smaller mean number of barks when displaying at their mutual boundary (see Figure 8.10), the differences were not statistically significant. Of considerable importance, however, was the finding that for both males the repetition rate of a series of barks was significantly more rapid when displaying at close range on their mutual boundary than when they were in the center of their territories (P's $< .05$).

Results of the quantitative analysis of *Zalophus* temporal barking patterns during the 1969 observations of an all-male captive group are clearly consistent with the impression that Peterson and Bartholomew obtained in their field study, i.e., that rapid barking tends to occur when there is close-range social interaction between bulls, and that the number of barks in a series is generally greater when the bull is directing his vocal threats at a specific individual while he is locomoting (or patrolling a territory) than when he is routinely self-advertising or stationary. Moreover, the present results suggest that a series of barks is also more rapid when directed at a specific opponent than when the series is not specifically directed.

In the 1970 observations, only one of the two bulls (Growler) barked a great deal while the other (Marty) showed his submission by remaining virtually silent. Marty remained in one pool and was accompanied by a juvenile male (4-year-old), while Growler stayed in the other pool with all four females and another juvenile male (4-year-old).

While Marty, always silent, swam in a clockwise stereotyped fashion, raising his head periodically to check the activities of Growler in the second pool, Growler's most prominent activity was herding of the four females. Generally, Growler attempted to keep the females in the same pool with him. If any or all of the females escaped the general vicinity of "Growler's pool" and went into "Marty's pool," Growler invariably would enter Marty's pool, barking both in air and underwater and chase the females back to his pool. During these encounters, Marty would leave the pool, slink away with lowered head and with an open, relaxed mouth similar to the appeasement displays described by Sandegren (1972) for Steller sea lions.

Table 8.4 Analysis of Barking Patterns by California Sea Lions
Growler and Marty in Compound 1 During an 8-minute Observation Period
(Marty held pool and Growler held eastern portion of compound.
Boundary between territories was at edge of pool.)
(Date: 5/29/69   Time: 16:10)

	GROWLER				MARTY			
	No. Barks/Series		Barks/Sec.		No. Barks/Series		Barks/Sec.	
	$\bar{X}$	SD	$\bar{X}$	SD	$\bar{X}$	SD	$\bar{X}$	SD
Boundary	7.5	4.4	2.8	0.1	5.3	9.8	3.2	0.4
Nonboundary	13.4	7.4	2.5	0.3	9.9	4.9	2.4	0.1

	TOTAL NO. OF BARKING SERIES	TOTAL NO. OF BARKING SERIES
Boundary	26	20
Nonboundary	9	70
Total	35	90

Miller (1974) has described the herding behavior of New
Zealand fur seals (*Arctocephalus forsteri*) in some detail. The
herding behavior of all otariid species is similar in many ways. In
the captive situation with *Zalophus*, Growler barked almost continuously, usually placing himself at a position in the oval-shaped
pool which most readily blocked the female escape route into
"Marty's pool." But, as with other otariid species, Growler's
herding behavior was frequently ineffective in keeping females in
his pool. When he "lost" the females, however, he never had
difficulty chasing them back to his pool. In addition to blocking
the females with his body (he was never observed biting a female),
Growler's most effective technique was to chase and bark at the
females. Frequently when a female appeared to be leaving
Growler's pool, she stopped and returned as Growler, barking
continuously in her direction, approached.

If Growler was resting in a sitting position and barking in a
routinely self-advertising way (i.e., in a series of nondirectional
barks), the mean repetition rate was 1.8 barks per second. The
mean repetition rate was 2.1 barks per second if he was patrolling

an area but not chasing or confronting either females or males. The mean rate was 2.5 barks per second if he chased or confronted others. Since the barking rates in a series within each of these social contexts were relatively stereotyped (standard deviations ranged from 0.1 to 0.3), each of these rates was significantly different from each other (P's $<$ .01). When chasing or confronting, moreover, Growler might increase his rate of barks during a single emitted series without a noticeable pause, and the mean rate of these rapid barks was 3.0.

Figure 8.10 Territorial display by male California sea lions M and G. Both animals are barking, with whiskers placed in a forward direction.

The repetition rate of underwater barking was almost exactly half the rate for a series of aerial barks. For example, Growler's mean rate of barking was 1.0 when he was either stationary or swimming slowly underwater; but if he was swimming rapidly while chasing another sea lion, his mean barking rate would go as high as 1.4 barks per second. Since few or no bubbles are produced when *Zalophus* barks underwater, the animal is presumably recycling its air in order to emit its underwater barks, causing the barks of a series to diminish to half the "normal" rate.

Thus it appears that much information content is contained in the temporal pattern and directionality of *Zalophus* barks. Adult *Zalophus* vary both the number of barks in a bark series as well as the rhythm of their barks depending on the social context. The receiver of these calls knows, to some extent by the intensity of the call, whether it is being addressed by the sender and can also "read out" the motivational state of the sender. In this fashion, the barking animal can readily control the movements and behavior patterns of its neighbors.

AVAILABILITY OF SEA LION VOCALIZATIONS FOR CONDITIONING

As Marler (1963) has noted, some aspects of the acoustical structure of species-specific mammalian vocalizations develop within a relatively closed genetic program. But the idea that nonhuman vocalizations are clearly defined involuntary emotional responses which are refractory to reinforcement contingencies and are, therefore, not subject to the kinds of stimulus controls exercised over nonvocal (e.g., bar-pressing, key-pecking, and paddle-pushing) responses is simply untrue (for a review of the literature on vocal conditioning in animals see Salzinger 1973).

As this chapter emphasized throughout, the vocalizations of both sexes and all ages of many pinniped species play a prominent role in their social behavior, and their ability to communicate frequently demands rather refined and precise vocalization control. It seems likely, therefore, that maintenance of seal and sea lion vocal behavior from infancy to adulthood acquires instrumental value, and its probability of occurrence in the presence of certain stimulus configurations (probably auditory, visual, tactual, and possibly olfactory) may be modified by schedules of reinforcement. The conditionability of *Zalophus* vocalizations was initially demonstrated experimentally when Schusterman and Feinstein (1965) elicited bursts of underwater clicks in a two-year-old female, then brought the vocalization under control of a visual cue (the size of circular or triangular stimuli) with the use of food reinforcement.

CONDITIONING AND CONTROLLING
UNDERWATER BURSTS OF "CLICKS"

The reliability of conditioning the emission of click bursts by *Zalophus* and controlling the likelihood of their occurrence by means of visual cues and reinforcement contingencies was demonstrated by replicating the original study of Schusterman and Feinstein (1965).

In all four sea lions, click vocalizations were readily elicited by the "frustration" technique, i.e., by withholding fish while the animal worked at a target-pressing task, an act that hitherto had yielded a fish reward. This technique quickly led to the production of click bursts, which were promptly reinforced with a piece of fish. Next, a vocal response (see Figure 8.11) was brought under the control of the size of circular and triangular target stimuli. Vocali-

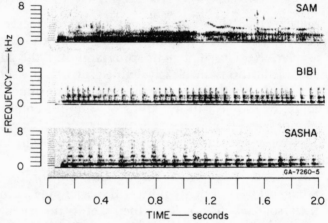

Figure 8.11 Spectrographs of conditioned underwater click bursts by one male and two female *Zalophus*.

zation in the presence of the large or small stimulus was reinforced, and silence in the presence of the opposite stimulus was also reinforced. Thus vocalization or silence in the presence of the appropriate stimuli defined the correct response. Acquisition curves for discriminative control of underwater click vocalizations in four sea lions are shown in Figure 8.12, which shows that discriminative control of click bursts by all four sea lions was consistently acquired within 600 total presentations; i.e., in 300 presentations of the stimulus controlling silence and 300 presentations of the stimulus controlling clicks. For two of the animals, in

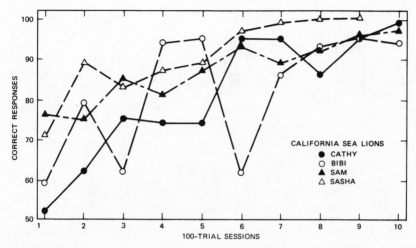

Figure 8.12 Acquisition curves for discriminative control of underwater click bursts. The cues were large and small circular discs.

fact, acquisition was relatively complete after only one session. Most errors during the early phase of discriminative control were "vocal" rather than "silent" errors; i.e., the sea lions frequently emitted a burst of clicks when presented with the stimulus calling for silence and only rarely remained silent when presented with the stimulus calling for click bursts. As conditioning proceeded, the opposite trend took place, and all animals showed some increase in the rate of silent errors during the final acquisition stages. The analysis of errors is shown in Figure 8.13.

Multiple stimulus control over the underwater click vocalizations of two sea lions (Sam and Bibi) was also obtained. Each sea lion emitted clicks underwater, indicating on each of twelve problems whether it could visually differentiate between two different stimulus shapes presented successively. One member of each pair of shapes was arbitrarily designated to require a vocalization, while the other member of the pair required silence. Each discrimination was learned to a criterion of twenty consecutive correct responses. Following the solution of these problems, over-

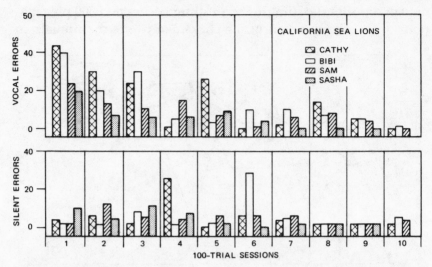

Figure 8.13 Error distributions based on the acquisition curves of Figure 8.12

learning trials were given and transfer tests were made in which the discriminative stimuli were reoriented. The results (see Table 8.5) showed that the California sea lion was capable of responding with a burst of underwater clicks when visually discriminating between several different geometric shapes, and that it was capable of transferring such discriminative vocal behavior when the shapes changed in orientation. In more recent work on discriminative control of vocal behavior, one sea lion (Bibi) was given the problem of associating its underwater clicks with one of two visual stimuli that differed in shape and size. The sea lion was presented with forty-five such problems, each of which lasted for 160 trials. Figure 8.14 shows the results of this experiment and clearly indicates that the sea lion learned to solve each of the discriminations. Furthermore, it showed improvement over problems, indicating that the conditioned vocal behavior of these animals is subject to the "learning-how-to-learn" phenomenon; i.e., a learning set can be formed during the acquisition of multiple stimulus control over vocal behavior.

Immediately following this demonstration, an attempt was made to determine how much visual information the sea lion Bibi could store and retrieve on a short-term basis when a vocal signal

Table 8.5 Multiple Stimulus Control of Sea Lion Vocalizations Under Water
(Performance of California sea lions, Bibi and Sam)

Stimulus[a] Pair No.	Original[b] Stimulus Shapes	Transfer Stimulus Shapes	Total Trials[c] To Learn		Total Errors		% Vocal Errors		Total Errors During 20 Overlearning Trials	
			Bibi	Sam	Bibi	Sam	Bibi	Sam	Bibi	Sam
1			180	72	71	21	43	43	0	1
2			135	140	63	71	81	62	0	0
3			221	251	93	75	66	61	2	5
4			62	294	20	86	85	41	0	1
5			66	256	12	124	92	40	1	1
6			272	139	132	48	60	85	2	1
7			67	126	17	42	100	38	1	2
8			235	761	115	385	81	46	2	4
9			428	555	215	213	73	58	4	2
10			115	37	33	5	52	25	0	0
11			118	23	38	3	74	100	4	1
12			20	199	2	86	100	42	0	1
Median			125.5	169.5	50.5	73.0	77.5	44.5	1.0	1.0
Range			20-428	23-761	2-215	3-385				

a. Significant positive transfer by Bibi (B) and Sam (S).
b. For Bibi, shape on left required silence and shape on right required vocalization; for Sam, the response requirements were reversed.
c. Includes criterion trials.

277

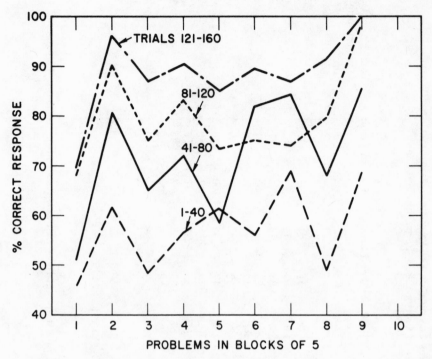

Figure 8.14 Interproblem learning curves for one California sea lion using conditioned vocalizations (click bursts) as an indicator response.

(click burst) was associated with one of two different geometric shapes. The technique used has been called serial discrimination learning or concurrent discrimination learning, and is similar to the serial learning of paired associates frequently used in the study of verbal learning with humans (Hayes, Thompson, and Hayes 1953). Instead of a given pair of shapes being repeatedly presented on successive trials until some criterion of learning has been achieved, or until some limited number of trials has been given, each pattern was presented for a single trial followed by another pattern and then a third, etc., through a list of ten patterns. The sea lion had to vocalize when it saw five of these patterns and had to remain silent when it saw five other patterns. To solve this problem, the sea lion would have to remember which stimulus required a vocalization and which stimulus required silence, and on any given run through a list of stimulus patterns the animal

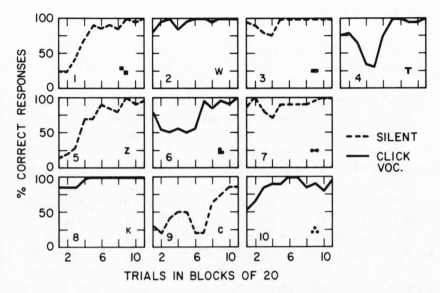

Figure 8.15 Acquisition curves for a concurrent discrimination learning problem in one California sea lion using conditioned vocalizations (click bursts) as an indicator response.

would have to remember all ten vocal-pattern associations. Figure 8.15 shows the results of this experiment and indicates that after approximately 160 trials, performance was maintained at better than 90 percent correct responses through the entire list of ten pattern discriminations.

In recent experiments, conditioned click vocalizations were used as an indicator or choice response to determine the California sea lion's visual acuity and auditory sensitivity (Schusterman and Balliet 1970; Schusterman, Balliet, and Nixon 1972; Schusterman 1974). Sea lions were trained to place their heads on a head holder so that all responses to acuity targets or to pulsed pure tones were made while the head was in a fixed position. Visual acuity thresholds obtained by the conditioned vocalization technique were even more reliable than those obtained by a paddle-pushing response and eliminated the difficulty of holding an animal in a fixed position as it was responding to the stimuli. Thus the use of conditioned vocalizations as an indicator response has proved extremely valuable in determining the California sea lion's differential sensitivity to both sights and sounds.

CONTROL OF BARKING

Andrew (1969) has been highly critical of previous research speculations dealing with the problems of modifiability and information content of emissions by odontocete whales and some pinniped forms. He suggests that there is currently less knowledge about the communication systems of these high-level marine mammals than there is for dogs, and that it is unfruitful to deal with these "sounds as a language whose phonemes have the same specificity of meaning as those in human speech." Support for this view has come from Bastian (1968), who reports that vocal signaling from one porpoise to another regarding a piece of arbitrary information emanating from the environment may be subject to a rather simple description in terms of operant conditioning procedures without invoking concepts such as "meaning," "language," and "syntax." The data previously presented show that manipulations of the environment or of social organization can result in previously neutral stimuli acquiring control over vocal emissions in *Zalophus*.

Even though conditioning probably plays a role in the vocal emissions of nonhuman primates (Sutton et al. 1973), attempts at conditioning vocalizations in several species have not been very successful (Myers 1969; Myers, Horel, and Pennypacker 1965; for a review, see Kellogg 1968). Furthermore, neurophysiological data suggest little or no extant evidence that the neocortex of nonhuman primates participates in the production or organization of their vocalizations (Myers 1969; Robinson 1967). On the other hand, the sound emissions of some odontocete whales (Turner 1962; Bastian 1967, 1968) and pinniped forms are readily conditionable and, in the case of the pinniped *Zalophus*, are subject to discriminative control by multiple stimuli. Perhaps then, for cetaceans and pinnipeds as well, a problem as fundamental as cataloging phonations or vocalizations and searching for the information content of these acoustic emissions is to determine the extent to which different acoustic emissions are available for conditioning as operant responses (Andrew 1969).

Although work of an experimental nature over a relatively wide range of species has demonstrated that the vocal behavior of nonhuman animals, e.g., emission rate, is as lawful as any other dimension of behavior and may be susceptible to variations of stimulus control and schedules of reinforcement (Warren 1965; Salzinger 1973), apparently little experimental work has been accomplished in the areas of animal vocalization and vocal communication aimed at problems in the modification of the topograph of vocalization. The success of behavior modification procedures across a wide range of behavioral repertoires in animals suggests that despite known biological constraints on learning, it is unwise at this time to postulate inherent neurophysiological limits of the organism as limiting the topography of its vocal repertoire.

As indicated earlier for *Zalophus*, information content is carried in the number of barks of a series as well as in their repetition rate. This is additional to other dimensions of their vocalizations, such as directionality, amplitude, and frequency range. Therefore, to further answer the question as to what extent different vocalizations in *Zalophus* are available for conditioning as operant responses, an attempt was made to condition different numbers of barks with food reinforcement.

The first step in training was to reinforce the sea lion (Sam) with a piece of fish every time it emitted a series of barks. As soon as barking in air came under the control of the trainer, the sea lion was trained to bark with its head underwater in a fixed position. In order to have a specific response which objectively indicated termination of barking, Sam was then trained to press a paddle subsequent to emitting barks underwater. Following this, the animal was reinforced with a piece of fish only when he barked in the presence of a 16-kilohertz tone.

An experiment to determine whether the number of barks in a series could be systematically changed was begun. Fish reinforcement was made contingent on the sea lion's emitting seven or more barks before pressing the paddle. The animal, its head in a fixed position, was given a two-second warning light (initiating a trial) followed by either a five-second tone or by "noise." The light and the tone terminated simultaneously. In a single test session, the sea lion was given approximately one hundred signal and one hundred noise trials which were presented according to a random sequence. Figure 8.16 shows the acquisition curve and indicates that a California sea lion can readily learn to bark a given number of times when given a simple auditory cue; the figure also shows that barks were rarely emitted on noise trials. Figure 8.17 shows the distribution of the number of barks for the entire sixteen test sessions and clearly shows that seven or eight barks were the most likely number to be emitted on a given signal trial.

In an experiment currently being conducted, the same sea lion has been trained to bark more than seven times in the presence of a loud acoustic signal and two to five times in the presence of a less intense signal. Figure 8.18 shows a spectrograph of the underwater barking responses to each of the signals. It should be noted that the repetition rate of this sea lion's conditioned underwater barking (with which bubbles were always associated) is considerably faster than heretofore reported in free-swimming *Zalophus*. Apparently, conditioning of the *number* of barks has also resulted in modification of the *rhythm* of barking. Thus the temporal dimension in *Zalophus* vocalizations, which is the primary carrier of information content (usually threat), is readily changed by reinforcement contingencies where the discriminative stimuli are pure tones and the reinforcer is food.

SUMMARY

Studies of captive and natural-living animals suggest that natural selection has favored the vocal-auditory communication channel in pinnipeds, particularly in those species having a polygynous breeding structure. Observations indicate that underwater as well as aerial calls function either to threaten or attract individuals by identifying the caller's species, sex, age, and location in space. Several lines of evidence suggest that mothers recognize vocalizations emitted by their own pups and that males recognize the phonations of other competing males on the breeding grounds.

The vocalizations of socially mature males (territorial or dominant) are extremely loud, directional, and repetitive. Female threat vocalizations, as well as pup-attraction calls, have been described as loud and intensive, and pup calls are also relatively loud and repetitive. The pup calls of several species contain harmonic structures with frequencies ranging from 0.2 to 8.0 kilohertz, and despite the fact that all pinniped species thus far tested show a hearing loss in air relative to their underwater

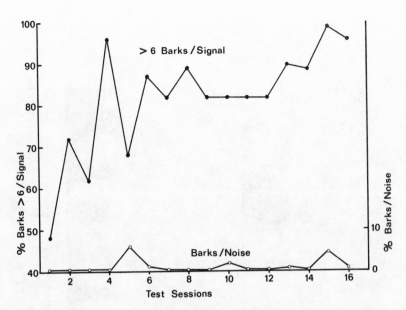

Figure 8.16 Acquisition curve for discriminative control over a specific number of underwater barks by a California sea lion. A pure tone was used as the discriminative stimulus.

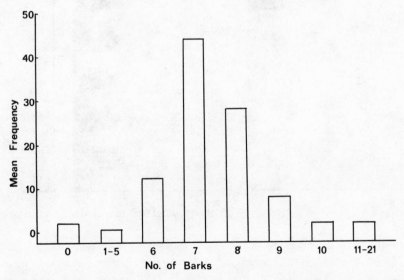

Figure 8.17 Distribution of the numbers of barks throughout sixteen test sessions of the conditioned barking experiment.

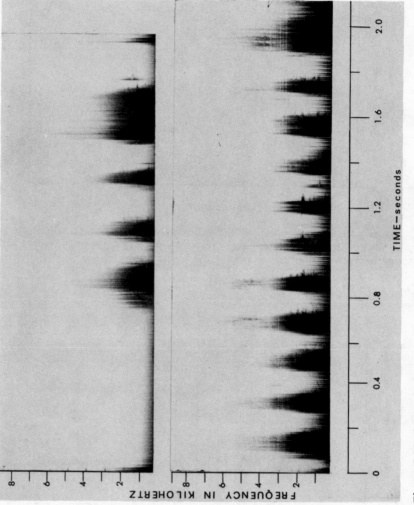

Figure 8.18 Spectrographs of conditioned underwater barks.

FREQUENCY IN KILOHERTZ

TIME—seconds

284

hearing, their hearing in air is well adapted for the reception of conspecific vocalizations (Schusterman 1974).

Both captive studies and observations in the wild show that the males of several pinniped species manifest seasonal variations in vocalizations. Quantitative studies on captive California sea lions show that they emit narrow cones of extremely loud aerial sounds, enabling them to address the loudest part of the signal toward the intended receiver. It is hypothesized that sea lion barking may play a significant role in female choice of a mate.

Although the "honking" vocalizations of male elephant seals during the breeding season have a typical intensity which does not vary with regard to social context, California sea lions grade their vocalizations depending on the social context. Quantitative studies show that sea lion bulls vary the number of barks in a series as well as the rhythm of their barks such that the recipient of these calls knows whether it is being addressed and can probably discriminate the motivational state of the sender.

Since pinniped ability to communicate frequently demands rather precise control of their vocalizations, it seems likely that the vocalizations of California sea lions may be readily brought under the control of a variety of stimulus configurations by means of reinforcement schedules. A variety of laboratory experiments have demonstrated that this is indeed true.

References

ABBREVIATION KEY

Am. Anthropol.	American Anthropologist
Am. J. Mental Deficiency	American Journal of Mental Deficiency
Amer. Psychol.	American Psychologist
Am. J. Phys. Anthropol.	American Journal of Physical Anthropology
Am. J. Physiol.	American Journal of Physiology
Am. J. Psychiatry	American Journal of Psychiatry
Am. Nat.	American Naturalist
Am. Psychol.	American Psychologist
Am. Sci.	American Scientist
Am. Zool.	American Zoologist
Anim. Behav.	Animal Behavior
Anim. Behav. Monogr.	Animal Behavior Monographs
Annu. Rev. Psychol.	Annual Review of Psychology
Antarct. J. U. S.	Antarctic Journal of the United States
Antarct. Res. Ser.	Antarctic Research Series
Arch. Neth. Zool.	Archives of Netherland Zoology
Aust. J. Zool.	Australian Journal of Zoology
Behav. Biol.	Behavioral Biology
Behav. Res. Methods Instrum.	Behavioral Research Methods and Instrumentation
Biol. Sci.	Biological Science
Biosci.	BioScience
Brain Res.	Brain Research
Can. Entomol.	Canadian Entomology
Can. Field Nat.	Canadian Field-Naturalist
Can. For. Zool.	Canadian Forestry and Zoology
Commun. Behav. Biol.	Communications in Behavioral Biology
Dev. Psychobio.	Developmental Psychobiology
Dev. Psychol.	Developmental Psychology
Ecol. Monogr.	Ecological Monographs

Folia Primatol.	Folia Primatologica
Harvey Lect.	Harvey Lectures Series
J. Accoust. Soc. Am.	Journal of the Acoustical Society of America
J. Anim. Ecol.	Journal of Animal Ecology
J. Appl. Behav. Anal.	Journal of Applied Behavior Analysis
J. Comp. Physiol. Psychol.	Journal of Comparative and Physiological Psychology
J. Ex. Anal. Behav.	Journal of the Experimental Analysis of Behavior
J. Exp. Psychol.	Journal of Experimental Psychology
J. Human Evol.	Journal of Human Evolution
J. Mammal.	Journal of Mammalogy
J. S. Afr. Wildl. Manage. Assoc.	Journal of the South African Wildlife Management Association
J. Theor. Biol.	Journal of Theoretical Biology
J. Wildl. Manage.	Journal of Wildlife Management
Nat. Geogr. Soc. Res. Rep.	National Geographic Society Research Reprints
Neuropsychol.	Nueropsychologia
Physiol. Behav.	Physiology and Behavior
Proc. R. Soc. London	Proceedings of the Royal Society of London
Psychol. Bull.	Psychological Bulletin
Psychol. Rec.	The Psychological Record
Psychol. Rep.	Psychological Reports
Psychol. Rev.	Psychological Review
Psychon. Sci.	Psychonomic Science
Psychosom. Res.	Psychosomatic Research
Q. Rev. of Biology	Quarterly Review of Biology
Radiat. Res. Suppl.	Radiation Research Supplement
Sci. Am.	Scientific American
Sci. Psychoanal.	Scientific Psychoanalysis
Soc. Biol.	Social Biology
U.S. Nat. Park Serv.	United States National Park Service
Z. Tierpsychol.	Zeitschrift fur Tierpsychologic

CHAPTER 1
CHENEY

Baum, W. 1974. Choice in free-ranging pigeons. *Science* 185:78– 79.

Beukemia, J.J. 1968. Predation by the three-spined stickleback (*Gasterorteus aculeatus L.*): The influence of hunger and experience. *Behaviour* 31:1– 126.

Bitterman, M.E. 1965. Phyletic differences in learning. *Am. Psychol.* 20:217-27.

Breland, K., and Breland, M. 1966. *Animal behavior.* New York: MacMillan.

Brower, L.P.; Cook, L.M.; and Croze, H.J. 1965. Predator response to artificial Batesian mimics released in neotropical environment. *Evolution* 21:11-23.

Burk, C.J. 1973. The kaibab deer incident: A long persisting myth. *Biol. Sci.* 23:113-14.

Chasan, D.J. 1974. Visitors learn in the Portland Zoo—but not more than monkeys, camels, ostriches, and giraffes. *Smithsonian* (July) 5:22-29.

Cheney, C.D. 1974. An ecological analysis of predation or, victims get what they ask for. Psi Chi address at Western Psychological Association meeting, April 26-28, San Francisco, California.

————, and Loether, R. 1975a. Proximity and activity as factors in prey selection by a great horned owl. Unpublished manuscript, Utah State University.

————. 1975b. Probability matching in four species of canids. Unpublished manuscript, Utah State University.

Cheney, C.D., and Snyder, R.L. 1974. A chamber for separating visual and physical access from predators. *Behav. Res. Methods Instrum.* 6:553-55.

————. Differential heartrate of a hawk viewing active or dead prey. *Condor*, in press.

Coppinger, R.P. 1970. The effect of experience and novelty on avian feeding behavior with reference to the evolution of warning coloration in butterflies: II. Reactions of naive birds to novel insects. *Am. Nat.* 104:323-35.

Denenberg, V.H.; Poschke, R.E.; and Zarrow, M.X. 1968. Killing of mice by rats prevented by early interaction between the two species. *Psychon. Sci.* 11:39.

Eaton, R.L. 1970. The predatory sequence, with emphasis on killing behavior and its ontogeny, in the cheetah (*Acinonyx jubatus* Schreber). *Z. Tierpsychol.* 27:492-504.

Eisenberg, J.F., and Leyhausen, P. 1972. The phylogenesis of predatory behavior in mammals. *Z. Tierpsychol.* 30:59-93.

Emlen, J.M. 1973. *Ecology: An evolutionary approach.* London: Addison-Wesley.

Erickson, A.W. 1972. Symposium on predator ecology and management. *J. Wildl. Manage.* 36:211-404.

Errington, P.L. 1963. The phenomenon of predation. *Am. Sci.* 51:180-92.

————. 1967. *Of predation and life.* Ames: Iowa State Univ. Press.

Ewer, R.F. 1973. *The carnivores.* Ithaca: Cornell Univ. Press.

Fox, M.W. 1969. Ontogeny of prey-killing behavior in Canidae. *Behaviour* 35:259-72.

Garcia, J.; Hankins, W.; and Rusiniak, K.W. 1974. Behavioral regulation of the milieu interne in man and rat. *Science* 185:824-31.

Hamilton, D.R. 1974. Immunosuppressive effects of predator induced stress in mice with acquired immunity to *Hymenolepis nana. J. Psychosom. Res.* 18:143-53.

Hardin, G. 1974. Lifeboat ethics: The case against helping the poor. *Psychology Today* (September) 8: 60-64.

Hemingway, E. 1932. *Death in the afternoon.* New York: Scribner.

Holling, C.S. 1959. The components of predation as revealed by a study of small mammal predation of the European pine sawfly. *Can. Entomol.* 91:293–320.

Hornocker, M.G. 1970. The American lion. *Natural History* (September) 79:40.

———. 1972. Predator ecology and management—what now? *J. Wildl. Manage.* 36:401–4.

Howard, W.E. 1974. *The biology of predator control.* Module in biology 11. London: Addison-Wesley.

Kaufman, D.W. 1973. Shrike prey selection: Color or conspicuousness? *Auk* 91:705–21.

———. 1974a. Differential owl predation on white and agouti *Mus musculus*. *Auk* 91:145–50.

———. 1974b. Differential predation on active and inactive prey by owls. *Auk* 91:172–73.

Krebs, J.R. 1973. Behavioral aspects of predation. In *Perspectives in ethology*, ed. P.P.G. Bateson and P.H. Klopfer, pp. 73–111. New York: Plenum Press.

Kruuk, H. 1972. *The spotted hyena—a study of predation and social behavior.* Chicago: Univ. of Chicago Press.

Linhart, S. 1974. Coyote research workshop. Paper read at Denver Wildlife Research Center, November.

Luckinbill, L.S. 1973. Coexistence in laboratory populations of *Paramecium aurelia* and its predator *Didinium nasutum*. *Ecology* 54:1320–27.

Luque, M. 1975. Ontogeny of fishing in Alaska brown bears. Master's thesis, Utah State University.

Markowitz, H. 1975. Analysis and control of behavior in the zoo. *Research in zoos and aquariums*. Washington, D.C.: National Academy of Sciences.

———. 1974. Servals—a report. Portland Zoological Society Newsletter (December) 3:2.

McInvaille, W.B., Jr., and Keith, L.B. 1974. Predator-prey relations and breeding biology of the great horned owl and red-tailed hawk in central Alberta. *Can. Field Nat.* 88:1–20.

Mech, L.D. 1970. *The wolf.* New York: Natural History Press.

———. 1974. A new profile for the wolf. *Natural History* (April) 83:26–31.

Metzgar, L.M. 1967. An experimental comparison of owl predation on resident and transient white-footed mice (*Peromyscus leucopos*). *J. Mammal.* 48:387–91.

Mueller, H.C. 1974. Factors influencing prey selection in the American kestrel. *Auk* 91:705–21.

Murton, R.K. 1967. Some predator-prey relationships in bird damage and population control. In *The problems of birds as pests*, ed. R.K. Murton and E.N. Wright, pp. 157–68. London: Academic Press.

Olsen, J. 1971. The poisoning of the west. *Sports Illustrated* 34: (March 8) pp. 80–93; (March 15) pp. 36–44; (March 22) pp. 34–42.

Parrish, J.D., and Saila, S.B. 1970. Interspecific competition, predation and species diversity. *J. Theor. Biol.* 27:207-20.

Pringle, L. 1975. Each antagonist in coyote debate is partly correct. *Smithsonian* (March) 6:74-81.

Royama, T. 1970. Factors governing the hunting behavior and selection of food by the great tit (*Parus major L.*). *J. Anim. Ecol.* 39:619-88.

Rudnai, J.A. 1973. *The social life of the lion.* Wallingford, Pa.: Washington Square East.

Ruggiero, L. 1975. The effect of interacting prey associated variables on prey selection by kestrels. Ph.D. dissertation, Utah State University.

———; Walls, S.; and Cheney, C. 1977. The experimental analysis of prey selection: a comment on methodology. *Psychol. Rep. J. Mammal.,* 40:555-558.

Schaller, G.B. 1969. Life with the king of beasts. *National Geographic* (April) 135:494-519.

Slobodkin, L.B. 1968. How to be a predator. *Am. Zool.* 8:43-51.

Snyder, R.L. 1974*a.* Probability matching in foxes. Paper read at Annual Animal Behavior Society meeting, June, at Champaign-Urbana, Illinois.

———. 1974*b.* Ferret and guinea pig performance on concurrent schedules. Paper read at Utah Academy of Science Arts and Letters meeting, April, Logan, Utah.

———. 1975. Some prey preference factors for a red-tailed hawk. *Auk* 92:547-52.

———, and Cheney, C. 1975*a.* Alternation in a two-choice prey selection experiment. *Condor,* in press.

———. 1975*b.* The predator: nature's conservationist. *Psychology Today,* in press.

———; Jensen, W.; and Cheney, C. 1976. Environmental familiarity and activity: aspects of prey selection for a ferruginous hawk. *Condor* 78:138-39.

Sparrowe, R.D. 1972. Prey-catching behavior in the sparrow hawk. *J. Wildl. Manage.* 36:297-308.

Tullock, G. 1971. The coal tit as a careful shopper. *Am. Nat.* 105:77-80.

Van Lawick, H., and Van Lawick-Goodall, J. 1970. *Innocent killers.* New York: Ballantine.

Van Lawick-Goodall, J. 1971. *In the shadow of man.* Boston: Houghton Mifflin.

Wagner, F.H. 1972. Coyotes and sheep. Paper read at 44th Faculty Honor Lecture, at Faculty Association, Utah State University, Logan, Utah.

Walls, S. 1975. The effects of dominance relationships in a coyote (*Canis latrans*) pack on dispersal. Ph.D. thesis, Utah State University.

CHAPTER 2
GUSTAVSON ET AL.

Allen, D.L. 1954. *Our wildlife legacy.* New York: Funk & Wagnalls.

Beidler, L.M. 1962. Taste receptor stimulation. *Progress in biophysics and biophysical chemistry.* New York: Pergamon Press.

Bolles, R.C. 1970. Species-specific defense reactions and avoidance learning. *Psychol. Rev.* 77:32–48.

Brower, L. 1969. Ecological chemistry. *Sci. Am.* 220:22–29.

————. 1971. Prey coloration and predatory behavior. In *Topics in the study of life,* pp. 360–79. New York: Harper & Row.

Brown, L., and Amadon, D. 1968. *Eagles, hawks and falcons of the world,* vol. 1. New York: McGraw-Hill.

Capretta, P.J., and Rawls, N.H. 1972. Establishment of flavor preferences in rats: Importance of early nursing and weaning experiences. Paper read at Annual American Psychological Ass. meeting, September, at Honolulu, Hawaii.

Clark, F.W. 1972. Influence of jackrabbit density on coyote population change. *J. Wildl. Manage. 36.*

Fitch, H.S.; Swenson, F.; and Tillotson, D.E. 1946. Behavior and food habits of the red-tailed hawk. *Condor* 48:205–57.

Galef, B.G., and Clark, M.M. 1971. Social factors in the poison avoidance and feeding behavior of wild and domesticated rat pups. *J. Comp. Physiol. Psychol.* 78:341–57.

————. 1972. Mother's milk and adult presence: Two factors determining initial dietary selection by weaning rats. *J. Comp. Physiol. Psychol.* 78:220–25.

Garcia, J.; Clark, J.C.; and Hankins, W. 1973. Natural responses to scheduled rewards. In *Perspectives in ethology,* ed. P. Bateson and P. Klopfer, pp. 1–41. New York: Plenum Press.

Garcia, J., and Ervin, F.R. 1968. Gustatory-visceral and telereceptor-cutaneous conditioning-adaptation in internal and external milieus. *Commun. Behav. Biol.* 1:389–415.

————; Yorke, C.; and Koelling, R. 1967. Conditionings with delayed vitamin injections. *Science* 1955:716–18.

————; and Koelling, R. 1966. Learning with prolonged delay of reinforcement. *Psychon. Sci.* 5:121–22.

Garcia, J., and Hankins, W. 1975. The evolution of bitter and the acquisition of toxiphobia. In *Fifth international symposium on olfaction and taste.* ed. D. Denton and J.P. Coghlan, pp. 39–45. New York: Academic Press.

————; Robinson, J.; and Vogt, J. 1972. Baitshyness: Tests of CS-US mediation. *Physiol. Behav.* 8:807–10.

————; and Rusiniak, K. 1974. Behavioral regulation of the milieu interne in man and rat. *Science* 185:828–31.

Garcia, J.; Kimmeldorf, D.; and Koelling, R. 1955. A conditioned aversion towards saccharin resulting from exposure to gamma radiation. *Science* 122:157–59.

Garcia, J., and Koelling, R. 1966. Relation of cue to consequence in avoidance learning. *Psychon. Sci.* 4:123–24.

———. 1967. A comparison of aversions induced by x-rays, toxins and drugs in the rat. *Radiat. Res. Suppl.* 7:439-50.

Garcia, J.; Kovner, R.; and Green, K.F. 1970. Cue properties vs. palatability of flavors in avoidance learning. *Psychon. Sci.* 20:313-14.

Green, K.F., and Garcia, J. 1971. Recuperation from illness: Flavor enhancement in rats. *Science* 1973:749-51.

Grossen, N.E. and Kelley, D. 1971. Role of species-specific responses and safety in the acquisition of avoidance behavior in rats. Paper read at Western Psychological Association meeting, April, San Francisco, California.

Gustavson, C.R. 1972. Response incompatibility between food reinforced key pecking and the UCR to shock in the pigeon. Master's thesis, University of Utah.

———. 1974. Taste aversion conditioning as a predator control method in the coyote and ferret. Ph.D. dissertation, Univ. of Utah.

———, and Garcia, J. 1974. Pulling a gag on the wily coyote. *Psychology Today* (August) 8:68-72.

———; Hankins, W.; and Rusiniak, K. 1974. Coyote predation control by aversive conditioning. *Science* 184:581-83.

———, and Gustavson, J.C. 1974. Predator control: A solution beyond warfare. *Defenders of Wildlife News,* August.

Hall, F.C. 1972. *Federal research. Natural areas in Oregon and Washington—A guide book for scientists and educators.* Portland: Pacific Northwest Forest and Range Experiment Station.

Hankins, W.; Garcia, J.; and Rusiniak, K. 1973. Dissociation of odor and taste in baitshyness. *Behav. Biol.* 8:407-19.

Harris, L.J.; Clay, J.; Hargreaves, F.; and Ward, A. 1933. Appetite and choice of diet: The ability of the vitamin B deficient rat to discriminate between diets containing and lacking the vitamin. *Proc. R. Soc. London* 113:162-90.

Holling, C.S. 1953. Sensory stimuli involved in the location and selection of sawfly cocoons by small mammals. *Can. For. Zool.* 36:633-53.

———. 1959. The components of predation as revealed by a study of the small-mammal predation of the European pine sawfly. *Can. Entom.* 91:293-320.

Kelly, Daniel J. Naturalistic observation of copulatory behavior in free-ranging coyotes. Unpublished manuscript, Eastern Washington State College.

Kohn, A. J. 1959. The ecology of *Conus* in Hawaii. *Ecol. Monogr.* 29:47-90.

Linhart, S.B.; Brusman, H.H.; and Balsar, D.S. 1968. Field evaluation of an antifertility agent, stilbestrol, for inhibiting coyote reproduction. *Trans. 33rd North Am. Wildl. Nat. Resour. Conf.* 316-27.

Marler, P., and Hamilton W., eds. 1967. *Mechanisms of animal behavior.* New York: Wiley.

Mayr, E. *Animal species and evolution.* 1963. Cambridge: Harvard Univ. Press.

———. *Principles of systematic zoology.* 1969. San Francisco: McGraw-Hill.

Nachman, M. 1963. Learned aversion to taste of lithium chloride and generalization to other salts. *J. Comp. Physiol. Psychol.* 56:343-49.

Prop, N. 1960. Protection against birds and parasites to some species of tenthredinid larvae. *Arch. Neth. Zool.* 13:380-447.

Revusky, S., and Garcia, J. 1970. Learned associations over long delays. In *Psychology of learning and motivation: Advances in research and theory 4,* ed. G. Bower and J. Spence. New York: Academic Press.

Richter, C.P. 1943. Total self-regulatory functions in animals and human beings. *Harvey Lect.* 38:53-103.

————; Holt, L.E., Jr.; Barelare, B., Jr.; and Hawkes, C.D. 1938. Changes in fat, carbohydrate and protein appetite in vitamin B deficiency. *Am. J. Physiol.* 124:596-602.

Rozin, P. 1967. Specific aversions as a component of specific hungers. *J. Comp. Physiol. Psychol.* 64:237-42.

————. 1968. Specific aversions and neophobia resulting from vitamin deficiency or poisoning in half-wild and domestic rats. *J. Comp. Physiol. Psychol.* 66:82-88.

————. 1969. Adaptive food sampling patterns in vitamin deficient rats. *J. Comp. Physiol. Psychol.* 69:126-32.

Ruiter, L. de. 1952. Some experiments on the camouflage of stick caterpillars. *Behaviour* 4:222-32.

————. 1956. Countershading in caterpillars: An analysis of its adaptive significance. *Arch. Neth. Zool.* 2:285-341.

Rusiniak, K.W.; Gustavson, C.R.; Hankins, W. G.; and Garcia, J. 1976. Prey-lithium aversions. II. Laboratory rats and ferrets. *Behav. Biol.* 17:73-85.

Scott, J.P. 1967. The effects of early experience on social behavior and organization. In *Social behavior and organization among vertebrates.* ed. W. Etkin, pp. 231-55. Chicago: Univ. of Chicago Press.

Smith, R.F.; Gustavson, C.R.; and Gregor, G.L. 1972. Incompatibility between the pigeons' unconditioned response to shock and the conditioned keypeck response. *J. Exp. Anal. Behav.* 18:147-53.

Tinbergen, L. 1960. The natural control of insects in pinewoods: I. Factors influencing the intensity of predation by songbirds. *Arch. Neth. Zool.* 13:266-336.

Wagner, F. 1971. Predator-prey instability and diversity of the Curley Valley ecosystem. Paper read at Southwestern and Rocky Mountain Division meeting of American Association for the Advancement of Science, at Tempe, Arizona.

————. 1972. Coyotes and sheep. Paper read at 44th Faculty Honor Lecture, January 1972, at Faculty Association, Utah State University, Logan, Utah.

Wagner, R. 1944. *Nutritional difference in the Mulleri group.* Univ. of Texas Publ. no. 4920, 39-41.

Wilcoxin, H.C.; Dragoin, W.B.; and Dral, P.A. 1971. Illness-induced aversions in rat and quail: Relative salience of visual and gustatory cues. *Science* 171:826-28.

CHAPTER 3
FENTRESS ET AL.

Altmann, S.A. 1967. The structure of primate social communication. In *Social communication among primates*, ed. S.A. Altmann, pp. 325-62. Chicago: Univ. of Chicago Press.

Cowan, I.M. 1947. The timber wolf in the Rocky Mountain national parks of Canada. *Can. J. Res.* 25:139-74.

Criddle, S. 1947. Timber wolf den and pups. *Can. Field Nat.* 61:115.

Crisler, L. 1958. *Arctic wild.* New York, Harper.

Eisenberg, J.F. 1976. Communication mechanisms and social integration in the Black Spider Monkey, *Ateles fuscicpes robustus*, and related species. *Smithson. Contrib. Zool.*, #213 Washington, D.C.: Smithsonian Institution Press.

Fentress, J.C. 1967. Observation on the behavioral development of a hand-reared male timber wolf. *Am. Zool.* 7:339-51.

_____ . 1973. Specific and non-specific factors in the causation of behavior. In *Perspectives in ethology*, ed. P.P.G. Bateson and P.H. Klopfer, pp. 115-224. New York: Plenum Press.

_____ , ed. 1976a. *Simpler networks and behavior.* Sunderland: Sinauer Associates.

_____ . 1976b. Dynamic boundaries of patterned behaviour: Interaction and self-organization. In *Growing points in ethology*, ed. P. P. G. Bateson and R. A. Hinde, pp. 135-69. London: Cambridge Univ. Press.

_____ , and Field-Lockhard, R. 1974. *Communication networks and social structures.* Paper read at American Association for the Advancement of Science meeting, February 28, at San Francisco, California.

Field, R. 1975. A perspective on syntactics of wolf vocalizations. Paper read at the Symposium on Behavior and Ecology of Wolves, annual meeting of the Animal Behavior Society, Wilmington, North Carolina.

_____ . 1976. Application of a digitizer for measuring sound spectrograms. *Behav. Biol.* 17:579-83.

Gautier, J.P. 1974. Field and laboratory studies of the vocalizations of talapoin monkeys (*Miopithecus talapoin*). *Behaviour* 51:209-73.

Golani, I. 1976. Homeostatic motor processes in mammalian interactions—a choreography of display. In *Perspectives in ethology*, vol. 2, ed. P.P.G. Bateson and P.H. Klopfer, pp. 69-134. New York: Plenum Press.

Green, S. 1975. Variation of vocal pattern with social situation in the Japanese monkey (*Macaca fuscata*): A field study. In *Primate behavior*, vol. 4, ed. R.A. Rosenblum, pp. 2-102. New York: Academic Press.

Greenwalt, C.H. 1968. *Bird song: Acoustics and physiology.* Washington, D.C.: Smithsonian Institution.

Haber, G.C. 1968. The social structure and behavior of an Alaskan wolf population. M.A. thesis, Northern Michigan University, Marquette.

Harrington, F.H. 1975. *Response parameters of elicited wolf howling.* Ph.D. thesis, State University of New York.

Hinde, R.A. 1972. *Social behavior and its development in subhuman primates.* Eugene: Univ. of Oregon Press.

Hinde, R.A., and Stevenson-Hinde, J. 1976. Towards understanding relationships: dynamic stability. In *Growing points in ethology*, ed. P.P.G. Bateson and R.A. Hinde, pp. 451–80. London: Cambridge Univ. Press.

Jordan, P.A.; Shelton, P.C.; and Allen, D.L. 1967. Numbers, turnover and social structure of the Isle Royale wolf population. *Am. Zool.* 7:233–52.

Joslin, P.W.B. 1966. *Summer activities of two timber wolf (Canis lupus) packs in Algonquin Park.* Master's thesis, Univ. of Toronto.

———. 1967. Movements and homesites of timber wolves in Algonquin Park. *Am. Zool.* 7:179–88.

Klinghammer, E. 1975. Analysis of 14 months of daily howl records in a captive wolf pack. Paper read at the Animal Behavior Society meetings, June, in Wilmington, N.C.

Kramer, G. 1963. Biobachtungen am einem von uns aufgezogen Wolf. *Z. Tierpsychol.* 18:91–109.

Marler, P. 1965. Communication in monkeys and apes. In *Primate behavior: field studies of monkeys and apes*, ed. I. DeVore, pp. 544–84. New York: Holt, Rinehart & Winston.

———. 1972. Vocalizations of East African monkeys. II: Black and white colobus. *Behaviour* 42:175–97.

———. 1973. A comparison of vocalizations of red-tailed monkeys and blue monkeys, *Cercopithecus ascanius* and *C. mitis*, in Uganda. *Z. Tierpsychol.* 33:223–47.

Mech, L.D. 1966. *The wolves of Isle Royale.* U.S. Nat. Park Serv. Fauna Series No. 7.

———. 1970. *The wolf.* New York: Natural History Press.

Murie, A. 1944. *The wolves of Mount McKinley.* U.S. Nat. Park Serv. Fauna Series No. 5.

Peterson, R.O. 1974. *Wolf ecology and prey relationships on Isle Royale.* Ph.D. dissertation, Purdue University.

Pimlott, D.H. 1960. The use of tape-recorded wolf howls to locate timber wolves. Paper read at the Twenty-Second Midwest Wildlife Congress, May.

Rutter, R. J., and Pimlott, D.H. 1968. *The world of the wolf.* Philadelphia: Lippincott.

Ryon, J. 1977. Den digging and related behavior in a captive timber wolf pack. *J. Mammology* 58:87–89.

Schenkel, R. 1947. Ausdrucks-studien an Wolfen. *Behaviour* 1:81–129.

Shalter, M.D.; Fentress, J.C.; and Young, G.W. 1977. Determinants of response of wolf pups to auditory signals. *Behaviour*, in press.

Simpson, M.J.A. 1976. The growth of knowledge and mother-infant relationships. In *Growing points in ethology*, ed. P.P.G. Bates and R.A. Hinde, pp. 385–400. London: Cambridge Univ. Press.

Skead, D.M. 1974. Incidence of calling in the black-backed jackal. *J. S. Afr. Wildl. Manage. Assoc.* 3:28-29.

Smith, W.J. 1968. Message-meaning analysis. In *Animal communication: Techniques of study and results of research*, ed. T.A. Sebeck, pp. 44-60. Bloomington: Indiana Univ. Press.

Tembrock, G. 1970. Bioakustische Untersuchungen au Saugetiern des Berliner Tierparkes. *Milu.* 3:78-96.

Theberge, J.B. 1966. Howling as a means of communication in timber wolves (*Canis lupus*). Master's thesis, Univ. of Toronto.

_____ , and Falls, J.B. 1967. Howling as a means of communication in timber wolves. *Am. Zool.* 7:331-38.

Thorpe, W.H. 1963. *Learning and instinct in animals.* 2nd ed. London: Methuen.

Wilson, E.O. 1975. *Sociobiology: The new synthesis.* Cambridge: Harvard Univ. Press.

Young, S.P., and Goldman, E.A. 1944. *The wolves of North America.* New York: Dover.

_____ , and Jackson, H.H.T. 1951. *The clever coyote.* Harrisburg: Stackpole Company.

Zimen, E. 1972. *Wolfe und Konigspudel.* Munich: Piper.

Chapter 4
Markowitz and Woodworth

Altmann, S.A. 1965. Sociobiology of rhesus monkeys II: Stochastics of social communication. *J. Theor. Biol.* 8:490-522.

Bandura, M. 1974. The effects of an operant conditioning experiment on the social behavior of a captive group of diana monkeys. Paper read at Western Psychological Association meeting, April 1974, at San Francisco, California.

Beach, F.A. 1955. The descent of instinct. *Psychol. Rev.* 62:401-10.

Bermant, G. and Alcock, J. 1973. Perspectives on animal behavior. In *Perspectives on animal behavior*, ed. G. Bermant, pp. 1-47. Glenview: Scott, Foresman.

Bobbitt, R.A.; Gourevitch, V.P.; Miller, L.E.; and Jensen, J.D. 1969. Dynamics of social interactive behavior: A computerized procedure for analyzing trends, patterns and sequences. *Psychol. Bull.* 71:110-21.

Box, H., and Poor, A.G. 1974. A quantitative method for studying behavior in small groups of monkeys in captivity. *Primates* 15:101-5.

Breland, K., and Breland, M. 1966. *Animal behavior.* New York: MacMillan.

Carpenter, R.C. 1964. A field study in Siam of the behavior and social relations of the gibbon. In *Naturalistic behavior of nonhuman primates*, ed. R.C. Carpenter, pp. 145-271. University Park: Pennsylvania State Univ. Press.

Chamove, A.S. 1974. A new primate social behavior category system. *Primates* 15:85-99.

Chasan, D.Z. 1974. In this zoo, visitors learn, though no more than animals. *Smithsonian* (July) 5(4):22-29.

Eibl-Eibesfeldt, I. 1970. *Ethology—the biology of behavior.* New York: Holt, Rinehart & Winston.

Ellefson, J.D. 1967. *A natural history of gibbons in the Malay Peninsula.* Ann Arbor: University Microfilms.

Ewer, R.F. 1968. *Ethology of mammals.* New York: Plenum Press.

Hess, E.H. 1962. Ethology: An approach toward the complete analysis of behavior. In *New directions in psychology,* ed. R. Brown, E. Galanter, E.H. Hess, and G. Mandler, pp. 157–266. New York: Holt, Rinehart & Winston.

Hinde, R.A. 1959. Some recent trends in ethology. In *Psychology: A study of a science,* vol. 2, S. Koch, pp. 560–610. New York: McGraw-Hill.

——— , and Atkinson, S. 1970. Assessing the roles of social partners in maintaining mutual proximity as exemplified by mother-infant relations in rhesus monkeys. *Anim. Behav.* 18:169–76.

Hopf, S. 1972. Study of spontaneous behavior in squirrel monkey groups: Observation techniques, recording devices, numerical evaluation and reliability tests. *Folia Primatol.* 17:363–88.

Kellogg, W.N., and Kellogg, L.A. 1933. *The ape and the child: A study of environmental influence on early behavior.* New York: McGraw-Hill.

Kendler, H.H. 1965. "What is learned?"—a theoretical blind alley. In *Controversial issues in learning,* ed. H. Goldstein, D.L. Krantz, and J.D. Rains, pp. 20–30. New York: Appleton-Century-Crofts.

Lehrman, D.S. 1953. Problems raised by instinct theories. *Q. Rev. Biol.* 28:337–65.

Lockard, R.B. 1971. Reflections on the fall of comparative psychology: Is there a message for us all? *Am. Psychol.* 26:168–79.

Lorenz, K.Z. 1970. *Studies in animal and human behavior,* vol. 1. Cambridge: Harvard Univ. Press.

Markowitz, H. 1973. Biological and behavioral research with captive exotic animals. Paper read at American Psychological Association meeting, August 1973, at Montreal, Quebec.

——— . 1974. New methods for increasing activity in zoo animals: Some results and proposals for the future. *Centennial symposium on science and research,* pp. 151–62. Topeka: Hill's Division, Riviana Foods.

——— . 1975. Analysis and control of behavior in the zoo. *Research in zoos and aquariums,* pp. 77–90. Washington, D.C.: National Academy of Sciences.

——— , and Becker, C.J. 1969. Superiority of "maze-dull" animals on visual tasks in an automated maze. *Psychon. Sci.* 17:171–72.

Markowitz, H.; Schmidt, M.; Nadal, L.; and Squier, L. 1975. Do elephants ever forget? *J. Appl. Behav. Anal.* 8:333–35.

Morris, D. 1970. The function and causation of courtship ceremonies. In *Patterns of reproductive behavior,* ed. D. Morris, pp. 128–52. New York: McGraw-Hill.

Restle, F. 1965. Discrimination of cues in mazes: A resolution of the "place-vs.-response" question. In *Controversial issues in learning,* ed. H. Goldstein,

D.L. Krantz, and J.D. Rains, pp. 260-74. New York: Appleton-Century-Crofts.

Schmuckal, G. 1974. The effects of feeding techniques upon the behavior of captive servals (*Felis serval*). Bachelor's thesis, Reed College.

Soper, E.T., III. 1974. The social behavior of captive diana monkeys under temporally dispersed and temporally localized feeding conditions. Master's thesis, Pacific University.

Tinbergen, N. 1953. *Social behavior in animals*. New York: Wiley.

Tuttle, R. 1972. Functional and evolutionary biology of hylobatid hands and feet. In *Gibbon and siamang*, vol. 1, ed. D.M. Rumbaugh, pp. 137-206. New York: Karger.

Van Lawick-Goodall, J. 1968. The behavior of free-living chimpanzees in the Gombe Stream Reserve. *Anim. Behav. Monogr.* 1:161-311.

————. 1971. *In the shadow of man*. Boston: Houghton Mifflin.

Yerkes, R.M. 1925. *Almost human*. London: Jonathan Cope.

CHAPTER 5
MYERS

Caldwell, M.G., and Caldwell, D.K. 1967. Intraspecific transfer of information via the pulsed sound in captive *odontocete cetaceans*. In *Les systems sonars animaux*, ed. R.C. Busnel, pp. 885-88. Italy: Frascati.

————. 1970. *Further studies on audible vocalizations of the Amazon freshwater dolphin, Inia geoffrensis*. Contributions in Science 187. Los Angeles: Los Angeles County Museum of Natural History.

Craighead, F.C., and Craighead, J.J. 1963. Radio tracking of grizzly bears in Yellowstone National Park. *Nat. Geogr. Soc. Res. Rep.*, pp. 59-68.

————. 1965.Biotelemetry research with grizzly bears and elk in Yellowstone Park, Wyoming. *Nat. Geogr. Soc. Res. Rep.*, pp. 49-62.

————. 1972. Grizzly bear prehibernation and denning activity as detailed by radio-tracking. *Wildlife Monogr.*, (32).

Craighead, J.J., and Craighead, F.C. 1971. Grizzly bear-man relationship in Yellowstone Park. *Biosci.* 21(16):845-57.

Harris, C.J. 1968. *Otters, a study of the recent Lutrinae*. London: Weidenfeld & Nicholson.

Hornocker, M.G. 1962. Population characteristics and social and reproductive behavior of the grizzly bear in Yellowstone National Park. Master of science thesis. University of Montana.

Layne, J.N. 1958. Observations on freshwater dolphins in the upper Amazon. *J. Mammal.* 39:1-22.

————, and Caldwell, D.K. 1964. Behavior of the Amazon dolphin, *Inia geoffrensis* (Blainville), in captivity. *Zoologica* (*N.Y.*) 49:81-108.

Liers, E. 1961. Notes on the river otter (*Luta canadensis*). *J. Mammal.* 32:1-9.

Martinka, C.J. 1971. Status and management of grizzly bear in Glacier Park.

Trans. North Amer. Wildlife Natur. Resources Conf. 36:312- 22.

Mundy, K.R.D., and Elook, D.R. 1973. Background for managing grizzly bears in the national parks of Canada. *Can. Wildlife Service Rep.* (22):1- 35.

Penner, R.H., and Murchison, A.E. 1970. Experimentally demonstrated echolocation in the Amazon river porpoise, *Inia geoffrensis* (Blainville). *Proc. 7th Annu. Conf. Biol. Sonar and Diving Mammals,* pp. 1- 22.

Pryor, K. 1975. *Lads before the wind.* New York: Simon & Schuster.

————— ; Haag, R.; and O'Reilly, J. 1969. The creative porpoise: Training for novel behavior. *J. Exp. Anal. Behav.* 12:653- 61.

CHAPTER 6
ESSOCK AND RUMBAUGH

Beck, B. 1974. Baboons, chimpanzees, and tools. *J. Human Evol.* 3:509-16.

Bruner, J.S. 1972. The uses of immaturity. *Amer. Psychol.* 27:687-708.

Connolly, C.J. 1950. *External morphology of the primate brain.* Springfield: Thomas.

Davenport, R.K., Jr., and Rogers, C.M. 1968. Intellectual performance of differentially reared cimpanzees: I. Delayed response. *Am. J. Mental Defic.* 72:674-80.

————— . 1970a. Intermodal equivalence of stimuli in apes. *Science* 168:279-80.

————— . 1970b. Differential rearing of the chimpanzee: A project survey. In *The chimpanzee,* vol. 3, ed. G.H. Bourne, pp. 377-80. Basel: Karger.

————— . 1971. Perception of photographs by apes. *Behaviour* 39:318-20.

————— ; and Menzel, E.W. 1969. Intellectual performance of differentially reared chimpanzees: II. Discrimination learning set. *Am. J. Ment. Defic.* 73:963-69.

————— ; and Russell, I.S. 1973. Cross-modal perception in apes. *Neuropsychol.* 11:21-28.

————— ; and Rumbaugh, D.M. 1973. Long-term cognitive deficits in chimpanzees associated with early impoverished rearing. *Dev. Psychol.* 9:343-47.

Ettlinger, G. 1967. Analysis of cross-modal effects and their relationship to language. In *Brain mechanisms underlying speech and language,* ed. F.L. Darley and C.H. Millikan, pp. 53-60. New York: Grune & Stratton.

Fossy, D. 1972. Living with mountain gorillas. In *The marvels of animal behavior,* ed. T. Allen. Washington, D.C.: National Geographic Soc.

Fouts, R. 1974. Language: Origins, definitions and chimpanzees. *J. Human Evol.* 3:475-82.

————— ; Mellgren, R.; and Lemon, W. 1973. American sign language in the chimpanzee: Chimpanzee to chimpanzee communication. Paper read at Midwestern Psychological Association meeting, May 1973, at Chicago, Illinois.

Gallup, G. 1970. Chimpanzees: Self-recognition. *Science* 167:86-87.

————— , and McClure, M. 1971. Preference for mirror-image stimulation in differentially reared rhesus monkeys. *J. Comp. Physiol. Psychol.* 78:403-7.

––––– ; and Hill, S. 1971. Capacity for self-recognition in differentially reared chimpanzees. *Psychol. Rec.* 21:69-74.

Gardner, B.T., and Gardner, R.A. 1971. Two-way communication with an infant chimpanzee. In *Behavior of nonhuman primates*, ed. A.M. Schrier and F. Stollnitz, pp. 117-82. New York: Academic Press.

Gardner, R., and Gardner, B. 1969. Teaching sign language to a chimpanzee. *Science* 165:664-72.

Gill, T.V., and Rumbaugh, D.M. 1974a. Learning processes of bright and dull apes. *J. Ment. Defic.* 78:683-87.

––––– . 1974b. Mastery of naming skills by a chimpanzee. *J. Human Evol.* 3:483-92.

Harlow, H.F. 1949. The formation of learning sets. *Psychol. Rev.* 56:51-65.

––––– ; Harlow, M.K.; Schiltz, K.A.; and Mohr, D.J. 1971. The effects of early adverse and enriched environments on the learning ability of rhesus monkeys. In *Cognitive processes of nonhuman primates*, ed. L.E. Jarrad, New York: Academic Press.

Harlow, H.F.; Schiltz, K.A.; and Harlow, M.K. 1969. Effects of social isolation on the learning performance of rhesus monkeys. *Proc. 2d Int. Congr. Primatol.* 1, Basel: Karger.

Hayes, K., and Hayes, C. 1952. Imitation in a home raised chimpanzee. *J. Comp. Physiol. Psychol.* 45:450-59.

Hodos, W. 1970. Evolutionary interpretation of nueral and behavioral studies of living vertebrates. In *The nuerosciences: Second study program*, ed. F.O. Schmidt, New York: Rockefeller Univ. Press.

Hull, C.L. 1943. *Principles of behavior.* New York: Appleton-Century-Crofts.

Kawai, M. 1965. Newly acquired precultural behavior of the natural troupe of Japanese monkeys on Koshima Islet. *Primates* 6:1-30.

Kohler, W. 1925. *The mentality of apes.* London: Routledge & Kegan Paul.

Le Gros Clark, W.E. 1959. *The antecedents of man.* Edinburgh: Edinburgh Univ.

Lieberman, P.; Crelin, E.; and Klatt, P. 1972. Phonetic ability and related anatomy of the newborn and adult human, Neanderthal man, and the chimpanzee. *Am. Anthropol.* 74:287-307.

Mason, W.A. 1968. Scope and potential of primate research. *Sci. Psychoanal.* 12:101-18.

Menzel, E. 1970. Menzel reporting on spontaneous use of poles as ladders. *Delta Primate Report* (December):1-4.

––––– . 1973. Further observations on the use of ladders in a group of young chimpanzees. *Folia Primatol.* 19:450-557.

––––– ; Davenport, R.K., Jr.; and Rogers, C.M. 1970. The development of tool using in wild-born and restricted-reared chimpanzees. *Folia Primatol.* 12:273-83.

Miles, R.C. 1957. Learning-set formation in the squirrel monkey. *J. Comp. Physiol. Psychol.* 50:356-57.

Nissen, H.W. 1951. Phylogenetic comparison. In *Handbook of experimental psychology*, ed. S.S. Stevens, pp. 347-86. New York: Wiley.

Noback, C.R., and Moskowitz, N. 1963. The primate nervous system: Functional and structural aspects in phylogeny. In *Evolutionary and genetic biology of primates*, vol. 1, ed. J. Buettner-Janusch, pp. 178-85. New York: Academic Press.

Parker, C. 1974. The antecedents of man the manipulator. *J. Human Evol.* 3:493-500.

Premack, D. 1971. On the assessment of language competence in the chimpanzee. In *Behavior of nonhuman primates*, ed. A.M. Schrier and F. Stollnitz, pp. 186-228. New York: Academic Press.

Rogers, C.M., and Davenport, R.K., Jr. 1971. Intellectual performance of differentially reared chimpanzees. III. Oddity. *Am. J. Ment.Defic.* 95:526-30.

Rumbaugh, D.M. 1969. The transfer index: An alternative measure of learning set. *Proc. 2d Int. Congr. Primatol.* 1:267-72, Basel: Karger.

————. 1970. Learning skills of anthropoids. In *Primate behavior: Developments in field and laboratory research*, ed. L.A. Rosenblum, New York: Academic Press.

————. 1971. Evidence of qualitative differences in learning processes among primates. *J. Comp. Physiol. Psychol.* 76:250-55.

————, and Gill, T.V. 1973. The learning skills of great apes. *J. Human Evol.* 2:171-79.

————. The learning skills of the Rhesus monkey. In *The rhesus monkey*, ed., G.H. Bourne. Academic Press, in press.

————. 1974. Language, apes, and the apple which-is orange, please. Paper read at Fifth International Congress of Primatology, August 1974, at Nagoya, Japan.

————; and von Glasersfeld, E. 1973. Reading and sentence completion by a chimpanzee (*Pan*). *Science* 182:731-33.

————; and Wright, S.C. 1973. Readiness to attend to visual foreground cues. *J. Human Evol.* 2:181-88.

Rumbaugh, D.M., and McCormack, C. 1967. The learning skills of primates: A comparative study of apes and monkeys. In *Progress in primatology*, ed. D. Starke, R. Schneider, and H.J. Kahn. Stuttgart: Fischer.

Rumbaugh, D.M.; von Glasersfeld, E.; Warner, H.; Pisani, P.; Gill, T.V.; Brown, J.; and Bell, C. 1973. A computer-controlled language training system for investigating the language skills of young apes. *Behav. Res. Methods Instrum.* 5:355-62.

Sackett, G.P., and Ruppenthal, G.C. 1973. Induction of social behavior changes in macaques by monkeys, machines, and maturity. In *Western Washington symposium on social learning*, ed. P.J. Elich, pp. 99-120. Bellingham: Western Wash. State College Press.

Schrier, A.M. 1958. Comparison of two methods of investigating the effect of amount of reward on performance. *J. Comp. Physiol. Psychol.* 51:725-31.

————, and Harlow, H.F. 1956. Effect of amount of incentive on discrimination learning by monkeys. *J. Comp. Physiol. Psychol.* 49:117-22.

Schrier, A., and Stollnitz, F., eds. 1971. *Behavior of nonhuman primates.* New York: Academic Press.

Smith, S.B. 1973. Transfer index testing in children. Master's thesis, Georgia State University.

Spence, K.W. 1956. *Behavior theory and conditioning.* New Haven: Yale Univ. Press.

Suomi, S.J., and Harlow, H.F. 1972. Social rehabilitation of isolate-reared monkeys. *Dev. Psychol.* 6:487–96.

———; and Novack, M.A. 1974. Reversal of social deficits produced by isolation rearing in monkeys. *J. Human Evol.* 3:527–34.

Turner, C.H.; Davenport, R.K., Jr.; and Rogers, C.M. 1969. The effect of early deprivation on the social behavior of adolescent chimpanzees. *Am. J. Psychiatry* 125:85–90.

Van Lawick-Goodall, J. 1968. The behaviour of free-living chimpanzees in the Gombe Stream Reserve. *Anim. Behav. Monogr.* 1:161–311.

———. 1970. Tool-using in primates and other vertebrates. In *Advances in the study of behaviour*, vol. 3, ed. R.A. Hinde and E. Shaw. New York: Academic Press.

———. 1971. *In the shadow of man.* Boston: Houghton Mifflin.

Warren, J.M. 1974. Possibly unique characteristics of learning by primates. *J. Human Evol.* 3:445–53.

Washburn, S.L. 1973. The promise of primatology. *Am. J. Phys. Anthropol.* 38:177–82.

———, and McCown, E.R. 1972. Evolution of human behavior. *Soc. Biol.* 19:163–70.

Yerkes, R.M., and Yerkes, A.W. 1929. *The great ape: A study of anthropoid life.* New Haven: Yale Univ. Press.

CHAPTER 7
STEVENS

Backhaus, D. Von. 1959a. Experimentelle untersuchungen uber die sehscharfe und das farbsehen einiger huftiere. *Z. Tierpsychol.* 16:445–67.

———. 1959b. Experimentelle prufung des farbsehvermugens einer massaigiraffe (*Giraffa camelopardalis tippelschirchi matschie* 1898). *Z. Tierpsychol.* 16:468–77.

Baldwin, J.; Stevens, V.J.; and Markowitz, H. 1973. Operant conditioning procedure for ungulates in a zoo setting. *Ankus.* 5:17–19.

Bandura, A. 1969. *Principles of behavior modification.* New York: Holt.

Baum, W.M. 1974. Choice in free-ranging wild pigeons. *Science* 185:78–79.

Beecher, M.D. 1971. Operant conditioning in the bat *Phyllostomus bastatus. J. Exp. Anal. Behav.* 16:219–23.

Bitterman, M.E. 1964. An instrumental technique for the turtle. *J. Exp. Anal. Behav.* 7:189–90.

Catania, A.C. 1966. Concurrent operants. In *Operant behavior: Areas of research and application*, ed. W.K. Honig, pp. 213–270. New York: Appleton-Century-Crofts.

Ewer, R.F. 1971. The biology and behavior of a free-living population of black rats (*Rattus rattus*). *Anim. Behav. Monog.* 4:127-76.

Felton, M., and Lyon, D.O. 1966. The post-reinforcement pause. *J. Exp. Anal. Behav.* 9:131-34.

Ferster, C.B., and Skinner, B.F. 1957. *Schedules of reinforcement.* New York: Appleton-Century-Crofts.

Grossman, K.E. 1973. Continuous, fixed-ratio, and fixed-interval reinforcement in honey bees. *J. Exp. Anal. Behav.* 20:105-9.

Grott, R., and Neuringer, A. 1974. Group behavior of rats under schedules of reinforcement. *J. Exp. Anal. Behav.* 22:311-21.

Herrnstein, R.J. 1966. Superstition: A corollary of the principles of operant conditioning. In *Operant behavior: Areas of research and application,* ed. W.K. Honig, pp. 33-51. New York: Appleton-Century-Crofts.

———. 1970. On the law of effect. *J. Exp. Anal. Behav.* 13:243-66.

Jambor, N.K., and Stevens, V.J. 1975. The effects of non-contingent reinforcement on operant responding in humans. Paper read at Western Psychological Association meeting, April 1975, at Sacramento, California.

Killeen, P. 1975. On the temporal control of behavior. *Psychol. Rev.* 82:89-115.

Lachter, G.D.; Cole, B.K.; and Schoenfeld, W.N. 1971. Response rate under varying frequency of non-contingent reinforcement. *J. Exp. Anal. Behav.* 15:233-36.

MacClintock, D., and Mochi, V. 1973. *A natural history of giraffes.* New York: Scribner.

Markowitz, H. 1975. Analysis and control of behavior in the zoo. In *Research in zoos and aquariums,* pp. 77-90. Washington, D.C.: National Academy of Sciences.

———; Schmidt, M; Nadal, L; and Squier, L. 1975. Do elephants ever forget? *J. Appl. Behav. Anal.* 8:333-35.

Markowitz, H., and Stevens, V.J. 1974. Tutorial on behavioral research in the zoo setting. Paper read at Western Psychological Association meeting, April 1974, at San Francisco, California.

Martin, G.R. 1974. Color vision in the tawny owl (*Strix aluco*). *J. Comp. Physiol. Psychol.* 88:133-41.

Myers, R.D., and Mesker, D.C. 1960. Operant responding in a horse under several schedules of reinforcement. *J. Exp. Anal. Behav.* 3:161-64.

Neuringer, A. 1969. Animals respond for food in the presence of free food. *Science* 166:339-41.

———. 1970. Superstitious key pecking after three peck-produced reinforcements. *J. Exp. Anal. Behav.* 13:127-34.

Penney, J., and Neuringer, A. 1974. The interaction of schedule-induced polydipsia and escape. Paper read at Western Psychological Association meeting, April 1974, at San Francisco, California.

Premack, D.; Schaeffer, R.W.; and Hundt, A. 1964. Reinforcement of drinking by running: Effect of fixed ratio and reinforcement time. *J. Exp. Anal. Behav.* 7:91-96.

Rimm, D.C., and Masters, J.C. 1974. *Behavior therapy: Techniques and empirical findings*. New York: Academic Press.

Schwartz, B. 1973. Maintenance of key pecking by response-independent food presentation: The role of the modality of the signal for food. *J. Exp. Anal. Behav.* 20:17-22.

Skinner, B.F. 1948. "Superstition" in the pigeon. *J. Exp. Psychol.* 38:168-72.

Squier, L.H. 1964. Operant conditioning of the Indian elephant (*Elephas maximus*). Paper read at American Psychological Association meeting, September 1964, at Los Angeles, California.

Staddon, J.E.R., and Simmelhag, V. 1971. The superstition experiment: A re-examination of its implications for the principles of adaptive behavior. *Psychol. Rev.* 78:3-43.

Stevens, V.J.; Baldwin, J.; Markowitz, H.; and Schmuckal, G. 1973. Operant conditioning of large ungulates in a zoo setting. Paper read at Animal Behavior Society meeting, December 1973, at Houston, Texas.

Thompson, T. 1963. Visual reinforcement in Siamese fighting fish. *Science* 141:55-57.

Turner, R.N., and Norris, K.S. 1966. Discriminative echolocation in a porpoise. *J. Exp. Anal. Behav.* 9:535-44.

Van Lawick-Goodall, J. 1968. The behavior of free-living chimpanzees in the Gombe Stream Reserve. *Anim. Behav. Monogr.* 1:161-311.

Wenzel, B.M.; Baldwin, B.A.; and Tschirgi, R.D. 1960. Operant conditioning of goats. *J. Exp. Anal. Behav.* 7:263-66.

Zeiler, M.D. 1968. Fixed and variable schedules of response-independent reinforcement. *J. Exp. Anal. Behav.* 11:405-14.

Zimmerman, W. 1974. Operant analysis and ethological study of the captive ostrich (*Struthio camelus*). Bachelor's thesis, Reed College.

CHAPTER 8
SCHUSTERMAN

Andrew, R.J. 1969. Signals and responses: A review of *Animal Communication*, ed. T.A. Sebeok. *Science* 164:693-94.

Barlow, G.W. 1972. A paternal role for bulls of the Galopagos Islands sea lion. *Evolution* 26:307-8.

―――. 1974. Galopagos sea lions are paternal. *Evolution* 28:476-78.

Bartholomew, G.A. 1953. Behavioral factors affecting social structure in the Alaska fur seal. *Trans. 18th N. Am. Wildl. Conf.*: 481-502.

―――. 1970. A model for the evolution of pinniped polygamy. *Evolution* 24:546-59.

―――, and Collias, N.E. 1962. The role of vocalization in the social behavior of the northern elephant seal. *Anim. Behav.* 10:7-14.

Bastian, J. 1967. The transmission of arbitrary environmental information between bottlenose dolphins. In *Animal sonar systems*, vol. 2, ed. R. Busnel, pp. 803-73. Jouy-en-Josas: Laboratoire de Physiologie Acoustique.

————. 1968. Further investigation of the transmission of arbitrary environmental information between bottlenose dolphins. *Naval Undersea Center TP* 109.

Cox, C.R., and LeBoeuf, B.J. Female incitation of male competition: A mechanism in sexual selection. *Amer. Nat.*, in press.

Eisenberg, J.F. 1966. The social organization of mammals. *Handbuch der Zoologie* 8 (10/7), Liefrung 39.

Gentry, R.L. 1970. Social behavior of the Steller sea lion. Ph.D. dissertation, Univ. of California.

Hayes, K.J.; Thompson, R.; and Hayes, C. 1953. Concurrent discrimination learning in chimpanzees. *J. Comp. Physiol. Psychol.* 46:105-7.

Kaufmann, G.W.; Siniff, D.B.; and Reichle, R.A. 1972. Colony behavior of Weddell seals, *Leptonychotes weddelli*, at Hutton Cliffs, Antarctica. Symposium on the biology of the seal. Guelph: Univ. of Guelph.

Kellogg, W.N. 1968. Communication and language in the home-raised chimpanzee. *Science* 162:423-27.

King, J.E. 1964. *Seals of the world.* London: British Museum (Natural History).

Kinne, O. 1975a. *Marine ecology*, vol. 2, part 2. New York: Wiley.

————. 1975b. Orientation in space: Animals-mammals. In *Marine ecology*, vol. 2, part 2, ed. O. Kinne, pp. 709-916. New York: Wiley.

Kok, O.B. 1972. Breeding success and territorial behavior of male boat-tailed grackles. *Auk* 89:528-40.

Kooyman, G.L. 1968. An analysis of some behavioral and physiological characteristics related to diving in the Weddell seal. *Antarct. Res. Ser.* 11:227-61.

LeBoeuf, B.J. 1974. Male-male competition and reproductive success in elephant seals. *Am. Zool.* 14:163-76.

————, and Peterson, R.S. 1969. Dialects in elephant seals. *Science* 166:1654-56.

Marler, P.R. 1963. Inheritance and learning in the development of animal vocalizations. In *Acoustic behavior of animals*, ed. R.G. Busnel, pp. 228-43. Amsterdam: Elsevier.

Miller, E.H. 1974. Social behavior between adult male and female fur seals, *Arctocephalus forsteri* (Lesson) during the breeding season. *Aust. J. Zool.* 22:155-73.

Moore, P.W.B. 1975. Underwater localization of click and pulsed pure-tone signals by the California sea lion (*Zalophus californianus*). *J. Acoust. Soc. Am.* 57:406-10.

————, and Au, W.W.L. 1975. Underwater localization of pulsed pure tones by the California sea lion (*Zalophus californianus*). *J. Acoust. Soc. Am.* 58:721-27.

Morris, D. 1957. "Typical intensity" and its relation to the problem of ritualization. *Behaviour* 11:1-12.

Myers, R.E. 1969. Neurology of social communication in primates. In *Recent advances in primatology: Neurophysiology*, Basel: Karger.

Myers, S.A.; Horel, J.A.; and Pennypacker, H.S. 1965. Operant control of vocal behavior in the monkey, *Cebus albifrons*. *Psychon. Sci.* 3:389-90.

Peterson, R.S. 1968. Social behavior in pinnipeds with particular reference to the northern fur seal. In *The behavior and physiology of pinnipeds*, ed. R.J. Harrison, R.C. Hubbard, R.S. Peterson, C.E. Rice and R.J. Schusterman, pp. 3–53. New York: Appleton-Century-Crofts.

———, and Bartholomew, G.A. 1967. *The natural history and behavior of the California sea lion.* Stillwater: Amer. Soc. Mamm.

———. 1969. Airborne vocal communication in the California sea lion. *Zalophus californianus. Anim. Behav.* 17:17–24.

Petrinovich, L. 1974. Individual recognition of pup vocalization by northern elephant seal mothers. *Z. Tierpsychol.* 34:308–12.

Poulter, T.C. 1963. Sonar signals of the sea lion. *Science* 139:753–55.

———. Underwater vocalization and behavior of pinnipeds. In *The behavior and physiology of pinnipeds*, ed. R.J. Harrison, R.C. Hubbard, R.S. Peterson, C.E. Rice, and R.J. Schusterman, pp. 69–84. New York: Appleton-Century-Crofts.

Ray, C. 1967. Social behavior and acoustics of the Weddell seal. *Antarct. J.U.S.* 2:105–6.

Robinson, B.W. 1967. Vocalization evoked from forebrain in *Macaca mulatta. Physiol. Behav.* 2:345–54.

Salzinger, K. 1973. Animal communication. In *Comparative psychology: A modern survey*, ed. D.A. Dewsbury and D.A. Rethlingshafer, pp. 161–93. New York: McGraw-Hill.

Sandegren, F. 1972. Sexual-agonistic signalling and territoriality in the Steller sea lion (*Eumetopias jubatus*). *Symposium on the biology of the seal.* Guelph: Univ. of Guelph.

———. Agonistic behavior in the male northern elephant seal. *Behaviour*, in press.

Scheffer, V.B. 1958. *Seals, sea lions and walruses: A review of the Pinnipedia.* Stanford: Stanford Univ. Press.

Schevill, W.E., and Watkins, W.A. 1965. Underwater calls of *Leptonychotes* (Weddell seal). *Zoologica (N.Y.)* 50:45–47.

Schusterman, R.J. 1968. Experimental laboratory studies of pinniped behavior. In *The behavior and physiology of pinnipeds*, ed. R.J. Harrison, R.C. Hubbard, R.S. Peterson, C.E. Rice, and R.J. Schusterman, pp. 87–171. New York: Appleton-Century-Crofts.

———. 1972. Pinniped sensory perception. *Symposium on the biology of the seal.* Guelph: Univ. of Guelph.

———. 1974. Auditory sensitivity of a California sea lion to airborne sound. *J. Acoust. Soc. Am.* 56:1248–51.

———, and Balliet, R.F. 1969. Underwater barking by male sea lions (*Zalophus californianus*). *Nature London* 222:1179–81.

———. 1970. Conditioned vocalization as a technique for determining visual acuity thresholds in the sea lion. *Science* 169:498–501.

———; and Nixon, J. 1972. Underwater audiogram of the California sea lion by conditioned vocalization technique. *J. Exp. Anal. Behav.* 17:339–50.

———; and St. John, S. 1970. Vocal displays by the gray seal, the harbor seal and the Steller sea lion. *Psychon. Sci.* 18:303-5.

Schusterman, R.J., and Dawson, R.G. 1968. Barking, dominance, and territoriality in male sea lions. *Science* 160:434-36.

Schusterman, R.J., and Feinstein, S.H. 1965. Shaping and discriminative control of underwater click vocalizations in a California sea lion. *Science* 150:1743-44.

Schusterman, R.J., and Gentry, R.L. 1971. Development of a fatted male phenomenon in California sea lions. *Dev. Psychobiol.* 4:333-38.

_____ ; and Schmook, J. 1966. Underwater vocalization by sea lions: Social and mirror stimuli. *Science* 154:540-42.

Stirling, I. 1971. Studies on the behavior of the South Australian fur seal, *Arctocephalus forsteri* (Lesson). I. Annual cycle, postures and calls, and adult males during the breeding season. *Aust. J. Zool.* 19:243-66.

_____ . 1973. Vocalization in the ringed seal (*Phoca hispida*). *J. Fish. Res. Board Can.* 10:1592-94.

_____ , and Warneke, R.M. 1971. Implications of a comparison of the airborne vocalizations and some aspects of the behavior of the two Australian fur seals, *Arctocephalus Spp.*, on the evolution and present taxonomy of the genus. *Aust. J. Zool.* 19:227-41.

Sutton, D.; Larson, C.; Taylor, E.M.; and Lindeman, R.C. 1973. Vocalization in rhesus monkeys: Conditionability. *Brain Res.* 52:225-31.

Turner, R.N. 1962. *Operant control of the vocal behavior of a dolphin.* Ph.D. dissertation, Univ. of California.

Warren, J.M. 1965. Comparative psychology of learning. In *Ann. Rev. Psychol.* Palo Alto: Stanford Univ. Press.

Winn, H.E., and Schneider, J. Communication in sereniens, sea otters and pinnipeds. In *Animal communication*, vol. 2, ed. T.A. Sebeok. Forthcoming.

Index

(continued)

The Editors

Hal Markowitz, Ph.D., is director of the Oregon Zoological Research Center and holds professional posts in biology and psychology at the University of Oregon Medical School and Portland State University. His work has appeared in many journals and periodicals including *Smithsonian Magazine, Der Stern,* and *Psychology Today.*

Victor J. Stevens, Ph.D., is an associate scientist and director of research education at the Oregon Zoological Research Center, Portland Zoological Gardens.